VOLCANOES

VOLCANOES

THIRD EDITION

Robert Decker
and
Barbara Decker

W. H. FREEMAN AND COMPANY
New York

Cover photograph: Anak Krakatau, September 1979
© Katia Krafft, used with permission

Library of Congress Cataloging-in-Publication Data

Decker, Robert W. (Robert Wayne), 1927-
    Volcanoes / Robert Decker and Barbara Decker. -- 3rd ed.
        p.   cm.
    Includes bibliographical references (p. -) and index.
    ISBN 0-7167-2440-5 (trade : acid-free paper). -- ISBN
0-7167-3174-6 (text/CD-Rom : acid-free paper)
    1. Volcanoes.   I. Decker, Barbara, 1929-   . II. Title.
QE521.D32   1997
551.21--dc21   97-17746
                                                                                                          CIP

Printed in the United States of America

Third Printing, 2002

Contents

Preface

Volcanoes assail the senses. They are beautiful in repose and awesome in eruption; they hiss and roar; they smell of brimstone. Their heat warms, their fires consume; they are the homes of gods and goddesses.

Volcanoes are described in words and pictures, but they must be experienced to be known. Their roots reach deep inside the Earth; their products are scattered in the sky. Understanding volcanoes is an unconquered challenge. This book poses more questions than answers; such is the harvest of curiosity.

Any book about volcanoes is bound to be highly descriptive, but we have tried to look behind these spectacular phenomena and to emphasize the processes involved. Our book is written for anyone—from student to expert—who is interested in learning more about how volcanoes work.

The list of Sources at the end of each chapter reflects only part of our debt to the hundreds of students of volcanoes whose general knowledge of volcanic processes and products we have used. Three friends deserve our special thanks: James G. Moore of the U.S. Geological Survey, the late Allan Cox of Stanford University, and the late John Staples of W. H. Freeman and Company. Without their help and encouragement, the first edition of this book would not have been written.

In the sixteen years since *Volcanoes* was first published there has been a great increase in public interest and awareness about volcanism. More than 200 volcanoes have erupted since the first edition of this book, most more than once; 30,000 people were killed by those

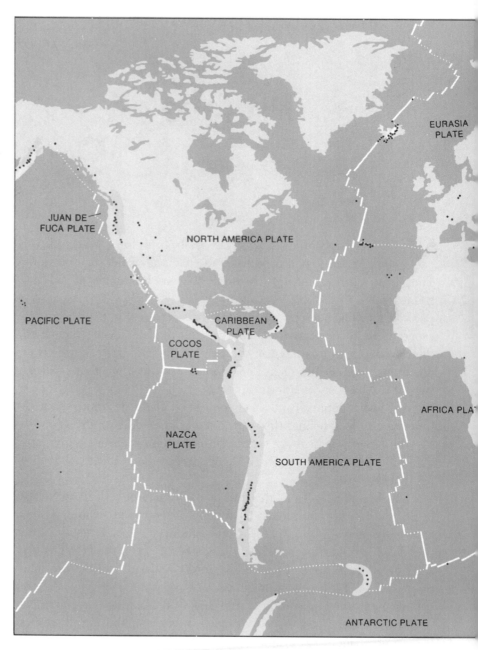

After Simkin, Tom, J. D. Unger, R. I. Tilling, P. R. Vogt, and H. Spall, *This Dynamic Planet: World map of volcanoes, earthquakes, impact craters, and plate tectonics*, U.S. Geological Survey, 1994.

EURASIA PLATE

PHILIPPINE PLATE

PACIFIC PLATE

AUSTRALIA PLATE

'RICA PLATE

ANTARCTIC PLATE

Key

Rift zones

Strike-slip (transform) faults

Subduction zones

· Geologically young volcanoes

eruptions. The fact that the most lethal eruptions during this period were from volcanoes not on our original list of "The World's 101 Most Notorious Volcanoes" is a sure sign that volcanology is still an evolving science.

But there has been good news during these years, too. Scientists have had some notable successes in forecasting explosive eruptions, particularly in the case of Pinatubo Volcano in the Philippines in 1991. Although more people have continued to crowd into cities and rich farmlands in the shadow of potentially dangerous volcanoes, during this same time scientists have gained a better understanding of volcanoes, and governing authorities have gained more respect for their potential danger. In the near future there will be close, life-or-death races between population pressure and successful eruption forecasting, with unknown outcomes. Most important, new ideas, new discoveries, and worldwide cooperation have helped geologists add more pieces to the puzzle of how volcanoes work.

Introduction

Most of the Earth's mountains, valleys, and plains have been slowly sculptured by uplift and erosion over the last million years, and these in turn have formed from other landscapes long vanished. Volcanoes, though, operate on a different time scale.

On the morning of May 18, 1980, Mount St. Helens in Washington State was a tall, conical volcano 2950 meters high; by nightfall it was an ugly stump of a mountain 2550 meters high, with a cloud of gas and ash rising from a new, horseshoe-shaped crater (Color Plates 15 and 16). A massive avalanche had torn away the whole north side of the mountain, unleashing a giant steam explosion that shattered the mountaintop, devastated a forest of 10 million trees, and killed 57 people. A huge eruption cloud jetted upward from the volcano, blanketing central Washington with ash and moving as far as the East Coast in only 3 days.

Volcanologists from all over the world converged on Washington State in a concerted effort to find answers to the specific questions of just what happened and why, as well as the larger questions of how the lessons learned at Mount St. Helens could be used for other dangerous volcanoes. Both applied and basic research benefited. Volcano-monitoring tools and techniques were refined, and the assessment of volcanic hazards in other potentially active areas was greatly improved.

The explosion of Mount St. Helens caught the rapt attention of the public too, when Americans who had thought of a volcanic eruption as something that happened in a Flintstones cartoon suddenly found volcanic ash sifting down into their backyards. They too wanted answers to such questions as: Why are volcanoes like Mount St. Helens explosive while others, like those in Hawaii, pour out quieter streams and fountains of lava? Why do volcanoes occur in chains like the Cascade Range or the Japanese islands? How does one know when a volcano is dead and not just dormant?

Mount St. Helens was still in the public consciousness when, 5 years later, another volcanic catastrophe shocked the world. In November 1985, Nevado del Ruiz Volcano in Colombia erupted, triggering mudflows that raced down its steep valley, engulfing a whole town and killing more than 20,000 people.

In the aftermath of this tragedy, scientists focused their attention on studying not just the details of the actual eruption, but also the problem of reducing volcanic risk by improved monitoring of hazardous volcanoes and by the use of more effective techniques for forecasting eruptions. Also of major concern was, and still is, the sensitive social and political problem of how and when to issue a forecast of a disaster that may or may not take place. Scientists do not want to be accused of crying "wolf"; the dilemma is that sometimes the wolf really is there but may not be hungry.

Forecasting the eruption of Pinatubo Volcano in the Philippines in 1991 was more successful. Although this giant explosive eruption—the second largest in this century—killed 300 people, 80,000 were evacuated from the most dangerous areas in the few days before the cataclysm. Fine volcanic ash and sulfuric acid aerosol droplets injected into the stratosphere by the Pinatubo eruption circulated worldwide and had a measurable effect on global climate.

Mount St. Helens, Nevado del Ruiz, and Pinatubo all are located on the Ring of Fire—the belt of young mountains, earthquakes, and volcanoes that encircles the Pacific Ocean. Why are volcanoes not randomly scattered about the Earth? We start with this question of why volcanoes are common in some regions of the Earth and not in others.

VOLCANOES

Seams of the Earth

1   San Andreas Fault cutting across the Elkhorn and Carrizo Plains in south central California. The North America Plate is on the right, and the Pacific Plate is on the left. Movement on the fault is called right lateral because for a person standing on either plate, the sense of motion on the opposite plate is to the right. (Photograph by Robert E. Wallace, U.S. Geological Survey)

Nature knows no pause in progress and development,
and attaches her curse to all inaction.

—Goethe (1749–1832)

he matching shores of eastern South America and western Africa form an intriguing jigsaw puzzle. Did Brazil's bulge once fit against the Congo? Did some great supercontinent break up and drift apart, each piece becoming one of our present continents?

Alfred Wegener, an Austrian scientist, championed the concept of continental drift for twenty years until his death on the Greenland Ice Cap in 1930. He noted not only that the edges of the continents make a rough fit but also that the Appalachian mountain range in eastern North America breaks off abruptly in Newfoundland and reappears across the Atlantic Ocean in Ireland, Scotland, and Scandinavia. He argued, too, that the similarity of European and American fossils up to 180 million years ago and the dissimilarity of more recent fossils is one more piece of evidence that the continents as we know them are still-drifting segments of an ancient supercontinent.

Wegener also recognized that the Earth's surface has two predominant elevations: a continental one between sea level and 1000 meters above sea level, and an ocean basin depth between 4000 and 5000 meters below sea level (Figure 2). He hypothesized that this difference in elevation between the continents and the oceans was caused by differences in the density and thickness of rocks in the continental blocks compared with that of the oceanic crust. He believed that the lighter and thicker continental blocks floated above the oceanic crust like icebergs drifting through the sea.

To understand Wegener's use of the terms *float* and *drift*, one must understand the great span of geologic time. Wegener envisioned the oceanic crust as being more viscous than tar or pitch, so that the "floating" took thousands of years to reach a balance and the "drifting" proceeded even more slowly. Continental drift was one of the great scientific controversies of its day, and Wegener's arguments in favor of it were alluring. Wegener called the great supercontinent he envisioned

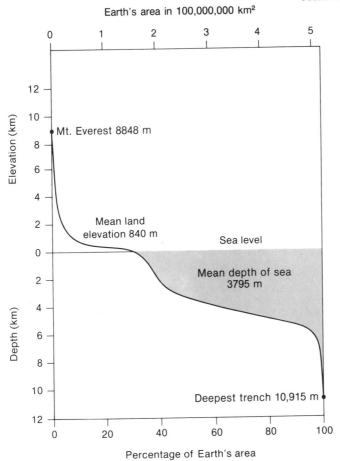

Earth's area in 100,000,000 km²

2   Graph comparing the Earth's surface elevations and surface areas. The two benches on the curve reflect the basically different thicknesses and densities of continental and oceanic crust. (Adapted from Alyn C. Duxbury, *The Earth and Its Oceans.* © 1971 Addison-Wesley, Reading, Massachusetts. Fig. 3.2. Reprinted with permission.)

Pangaea: it was so large that it contained both the polar scars of ancient ice ages and the equatorial legacy of vast coal swamps that existed before the continent's breakup. According to his theory, the present locations of glacial deposits near the equator and of coal deposits near the current poles are the result of continental drift.

But Wegener was also a zealot, and many of his contemporaries—wary of too much zeal—considered him a crackpot. In the first quarter of this century, geophysicists who specialized in the study of the

Earth's interior were especially skeptical because Wegener was unable to identify the strong forces needed to propel the continents away from one another. Because these experts had already concluded that the interior of the Earth was as solid and strong as steel, the notion that floating continents could drift through such a strong seafloor was untenable.

These geophysicists' knowledge of the deep interior of the Earth came mainly from analyzing what happens to earthquake waves as they propagate through the Earth, a branch of earth science called *seismology*. Its methods are similar to the doctor's technique of thumping on your chest to find out whether there is liquid or air in your lungs: the differences in the quality of the thump sound passing through your chest provide the key. Seismology is a rigorous discipline based on precise physical principles, and its conclusions are well respected by all geologists. The denial of continental drift because of such geophysical objections was convincing despite the good circumstantial evidence in its favor, and the hypothesis went underground for thirty years.

Among geologists, however, there is a half-joke called Smith's law, which states that anything that did happen, can happen. So even without understanding how the large horizontal movement of continents could happen, some geologists still looked for evidence that it did happen.

After World War II, seafloor topography was mapped by continuous echo sounders on oceanographic ships, and seismologists interested in the location of earthquakes plotted patterns of epicenters that coincided with the crests of the submarine ridges mapped by the oceanographers. Without openly declaring themselves, many earth scientists began to think that maybe Wegener was right.

In the fall of 1963, Harry Hess, a professor of geology at Princeton University, gave his presidential address to the Geological Society of America, which was entitled "Further Comments on the History of Ocean Basins." It was a bombshell. In his lecture, Hess built on a proposal made in 1928 by Arthur Holmes, an eminent British geologist, that the seafloor may be spreading apart (Figure 3). In Hess's elaboration, molten rock erupts from the Earth's interior along mid-ocean ridges, creating new seafloor that slowly spreads away from the ridges. To prevent this spreading from expanding the Earth, Hess suggested that the seafloor eventually sinks into the deep ocean trenches. With

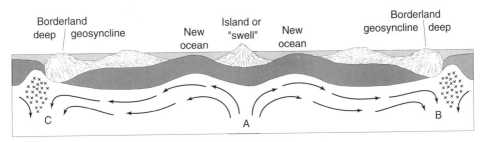

3 Original 1928 drawing by Arthur Holmes illustrating continental drift and the development of new ocean basins. Eclogite is a dense rock formed under high pressure that has the same general composition as basalt. In Holmes's own words: "A current ascending at A spreads out laterally, extends the continental block and drags the two main parts aside, provided that the obstruction of the old ocean floor can be overcome. This is accomplished by the formation of eclogite at B and C, where sub-continental currents meet sub-oceanic currents and turn downwards. Being heavy, the eclogite is carried down, so making room for the continents to advance." (From *Transactions of the Geological Society of Glasgow*, vol. 18, 1928–1929, p. 579. Reprinted in A. Holmes, *Principles of Physical Geology*, 2nd ed. New York, Ronald Press, 1965, p. 1001.)

Hess's address, Wegener's largely discredited ideas gained new legitimacy, and the earth science revolution began.

Evidence that helped prove Wegener and Hess right came in the early 1960s from marine geophysicists who were analyzing the Earth's magnetic field over the ocean floor. They found that the magnetic field was unusually strong directly over the oceanic ridges. The reason, they proposed, was that the rocks making up the ridge contained iron-rich minerals and were magnetized parallel to the Earth's magnetic field, which tends to reinforce that field.

As the geophysicists extended their surveys away from the ridges, they found that the magnetic field at the surface of the sea was a zebra-striped pattern of alternating belts of unusually high and unusually low intensity. Their explanation of the low-intensity belts was that the ocean floor beneath them had formed at times when the Earth's magnetic field pointed to the south, rather than to the north as it does today.

By the time of the marine magnetic surveys in the early 1960s, geophysicists studying the magnetic properties and ages of lavas on land had already demonstrated that the Earth's magnetic field periodically switches from north to south and back again. The time between reversals may be as short as 30,000 years or as long as several million years. Using this magnetic reversal time scale, the marine geophysicists knew when the field had flipped, and so they could estimate the age of the ocean floor beneath each magnetic stripe. The stage was set for a dramatic new idea.

Geologists now proposed that instead of the continents drifting through the ocean floor, the seafloor spread away from the oceanic ridges as it formed, carrying along the continents as part of large, spreading plates (Figure 4). This new idea was called *seafloor spreading* or *plate tectonics*, and the magnetic stripes provided a record of both the geometry and the rate of spreading, or the plate movements. The pattern and ages of the magnetic stripes parallel to the ocean's ridge systems provide a giant slow-motion tape recording of the Earth's horizontal movements over the past 170 million years.

The pattern that has emerged shows the Earth's surface broken into eleven large and five smaller plates that are moving in various directions relative to one another (Figure 5). The margins or seams between the plates are of three basic types: *convergent*, in which one plate overrides the other; *divergent*, in which the plates separate and move apart; and *side slipping* or *strike slip*, in which the plates slide by each other without separating or overriding (Figure 6). Convergent and divergent margins are characterized by different types of major volcanic activity, but strike-slip margins have little or no volcanism, so the link between plate tectonics and volcanoes is close indeed.

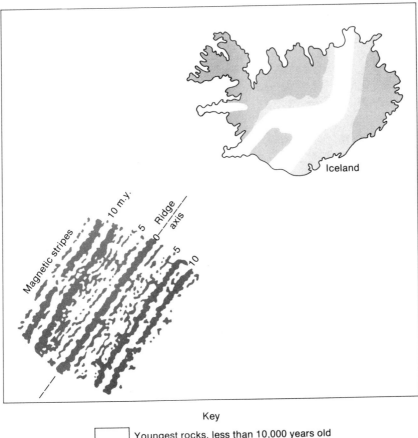

Key

Youngest rocks, less than 10,000 years old

Rocks 10,000 to 4,000,000 years old

Rocks 4,000,000 to 15,000,000 years old

m.y.   Millions of years

4   Magnetic field strength measured at the sea surface southwest of Iceland shows a definite striped pattern parallel to the axis of the Mid-Atlantic Ridge. The symmetry of the pattern on either side of the ridge and the similar pattern in the ages of volcanic rocks in Iceland are strong evidence that the seafloor spreads away from the ridge axis. (Adapted from Charles Drake, "The Geological Revolution," Condon Lectures, Oregon State System of Higher Education, Eugene, 1970.)

Key

— Subduction zone
— — — Uncertain plate boundary
— Strike-slip (transform) faults
→ Direction of plate motion
— Ridge axis

5 Rigid plates of the Earth's surface are slowly moving horizontally away from and toward each other. The arrows shown assume that the Africa Plate is not moving. Plates separate along the crests of mid-ocean ridges, slide past each other along strike-slip faults, and converge at subduction zones. (Adapted from J. F. Dewey, "Plate Tectonics." Copyright © 1972 by Scientific American, Inc. All rights reserved.)

Labels on map:

EURASIA PLATE
Caucasus
Carpathians
Alps
Anatolian fault
Carlsberg ridge
AFRICA PLATE
Atlantic-Indian ridge
ANTARCTIC PLATE
Reykjanes ridge
Mid-Atlantic ridge
NORTH AMERICA PLATE
CARIBBEAN PLATE
SOUTH AMERICA PLATE
Andes
Peru-Chile trench
Chile rise
NAZCA PLATE
COCOS PLATE
Mexico trench
East pacific rise
Rockies
San Andreas fault
JUAN DE FUCA PLATE
Aleutian trench
Kuril trench
Japan trench
Marianas trench
PACIFIC PLATE
New Hebrides trench
Kermadec-Tonga trench
Macquarie ridge
Pacific-Antarctic ridge
ANTARCTIC PLATE
PHILIPPINE PLATE
Java trench
AUSTRALIA PLATE
South East Indian rise
EURASIA PLATE
Himalayas

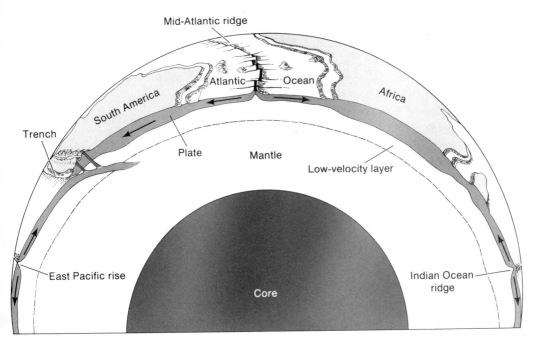

6 Movement of plates shown in a cross section of the Earth. The Africa and South America Plates separate along the Mid-Atlantic Ridge at a rate of a few inches per year. The South America and Nazca (east Pacific Ocean) Plates converge to form the Andes Mountains. The thickness of the plates and the low-velocity layer on which they move are exaggerated so that they can be shown at this small scale. (Adapted from K. C. Burke and J. Tuzo Wilson, "Hot Spots on the Earth's Surface." Copyright © 1976 by Scientific American, Inc. All rights reserved.)

Earthquakes, topography, and other geological structures identify the three types of seams. At the convergent margins, also known as *sub-duction zones*, one plate (generally the oceanic one), slips beneath the other. This causes earthquakes, which originate either near the surface seam or in the sinking plate at depths as great as 700 kilometers. Physical forces at the convergent margin cause the ocean floor to bend down into a deep trench while the edge of the upper plate is pushed into folds or is broken by *thrust faults* (in which the upper side moves up and over the lower side). Rock layers scraped off the ocean floor are stacked into slices or folded against the upper plate. The geological result is an island arc like Japan or a mountain chain like the Andes.

At divergent margins, the earthquakes are shallow and follow the center of the rift. The center of the rift is often a valley formed by the separating plates (Figure 7). No folds are formed because the area is being stretched, not squeezed, and the fractures are either open cracks or slumps called *normal faults*. Most divergent margins are submarine, and new oceanic crust is formed at the separating edges, with the crust spreading outward from the oceanic ridges near the centers of the world's major oceans. The main ridges are the Mid-Atlantic Ridge, the East Pacific Rise, and the Indian Ocean Ridges. These are broad, rugged, submarine mountain ranges a few kilometers high, a few thousand kilometers wide, and tens of thousands of kilometers long. Drain away the oceans, and you would be able to see the greatest mountain system on Earth.

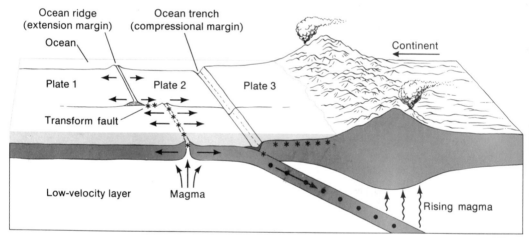

Key

* Shallow earthquakes
  (tension on ridges; lateral slip on transform faults)

• Deep earthquakes
  (mainly showing thrusting and down-dip compression)

7   A schematic cross section of plate margins showing the association of earthquakes with the plate boundaries formed by ocean ridges, strike-slip (transform) faults, and subduction zones. (Adapted from L. Sykes et al., in *Earth*, First Edition. F. Press and R. Siever, p. 644. W. H. Freeman and Company, Copyright © 1974.)

The strike-slip plate margins (named *transform faults* in the jargon of plate tectonics) are also mainly submarine. They connect offsets of the oceanic ridges into a rectangular pattern that, on topographical maps of the seafloor, looks like an alligator's back (Figure 8).

Sometimes a slice of continent gets entangled in the shearing of a side-slipping margin. The San Andreas Fault in California is a classic example of a transform fault on land. Los Angeles is moving northwest relative to San Francisco at a rate of about 5 centimeters per year. Even so, society will probably make them one city long before nature slides the two together. Shallow earthquakes, linear valleys, and twisted rocks mark the strike-slip plate margins (Figure 1).

Despite the overwhelming evidence that seafloor spreading takes place, the strength and solidity of the Earth's interior remained a problem that was finally resolved by seismologists. This is ironic, since it was the seismologists whose arguments had shot down Wegener's hypothesis in the 1920s. When they looked closely at earthquake wave records, particularly those from underground nuclear bomb tests with precisely known locations and times of origin, the seismologists discovered a zone about 100 to 200 kilometers beneath the Earth's surface where seismic waves travel unusually slowly and are partially absorbed.

This low-velocity layer is not very strong because it probably contains a small percentage of molten rock. The plates move on the soft, low-velocity layer under the influence of small horizontal forces, and fractures in the Earth's crust at the divergent plate margins allow the molten rock to leak to the surface, forming volcanoes (Figure 9).

What are the great forces that slowly move the plates? The options are simple—push, pull, slide, or drag—but the arguments are more complex. All these forces or, more precisely, *stresses* (forces divided by the areas on which they act), have their advocates, whose general arguments are as follows:

Geologists who vote for "push" envision a prying apart of the plates at their diverging margins. They suggest that the injection of magma into the cracks between the plates pushes them apart. Those who vote for "pull" believe that the subducting part of a plate probably is more dense than the rocks in the upper mantle that surround it, causing it to sink and pulling down the rest of the plate behind it. The problem with both push and pull is that they require the plate to be strong

8   *(Facing page)* A topographical map of the Atlantic Ocean with the water drained away. The Mid-Atlantic Ridge is one of the world's greatest mountain ranges. The breaks perpendicular to the valley crest of the ridge are called *fracture zones* or *transform faults*—other terms for strike-slip margins. (From *World Ocean Floor Panorama* by Bruce C. Heezen and Marie Tharp, 1977. Reproduced by permission of Marie Tharp, all rights reserved.)

enough to be moved by forces acting on their edges. But because plates are huge slabs of the Earth's crust and upper mantle, their very size makes them inherently weak. Opponents of push or pull use the well-known rule that to make a small-scale model behave realistically, you must reduce its strength proportional to the change in scale. So the opponents propose a thought experiment: Squeeze out a line of toothpaste on a sink top and then try to move the toothpaste tube by pushing

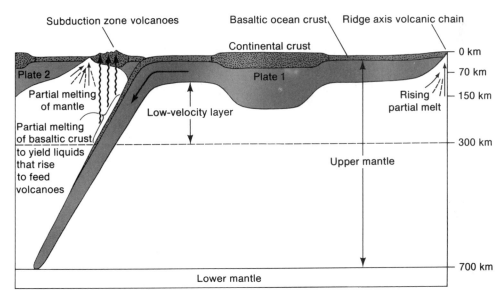

9   A cross section of the Earth's upper mantle. The rigid plate is composed of solidified rock that moves on the partly molten, low-velocity layer. The plates are approximately 70 kilometers thick under the oceans and 100 to 150 kilometers thick under the continents. The continents are part of the plates and so move with them. (Adapted from J. F. Dewey, "Plate Tectonics." Copyright © 1972 by Scientific American, Inc. All rights reserved.)

or pulling on the end of the line of paste. The proponents of push or pull counter that if you make the sink top slippery enough and push or pull slowly enough, it might work.

In regard to the idea of sliding plates, some geologists argue that gravity moves the plates from locations generally higher (the mid-ocean ridge axes) to locations generally lower (the ocean trenches). Gravity works internally on the whole plate—a so-called body force—and thus avoids the strength problem posed by forces acting only on or near the plate margins. According to this idea, a plate literally slides downhill from ridge to trench.

Dragging a plate by the flow of plastic rock in the upper mantle beneath it—the original suggestion of Arthur Holmes—also avoids the strength problem but does require some frictional connection between the bottom of the plate and the moving mantle material. Whereas gravity sliding requires a very slippery contact zone beneath a plate, dragging by currents slowly stirring in the upper mantle beneath a plate requires a more viscous zone.

Possibly the plates are propelled by a combination of stresses, one more dominant in some regions, another in other places. For now, the question of what exact stress moves a particular plate has not, in our opinion, been clearly answered, but unanswered questions are the lifeblood of scientific research.

Wegener turned out to be a prophet, although his original idea has been significantly changed. He believed that the continents drifted like isolated rafts through a viscous oceanic crust. The theory of seafloor spreading envisions rigid plates of oceanic crust forming and spreading from the ocean ridges, with the continental blocks carried like rafts frozen into the larger plates, which are composed of both oceanic and continental crust.

In the last four decades of unprecedented progress in understanding the Earth, the development and rapid acceptance of the concept of seafloor spreading stands as one of the great revolutions in science. Not only can the jigsaw puzzle of the continents be reconstructed, but their ongoing movements also can be determined. The dynamics of plate motions control both earthquakes and volcanoes, which occur, for the most part, on the creaking and leaking margins of the plates.

The association between volcanoes and earthquake belts has long been recognized. Aristotle proposed that earthquakes were caused by the rumbling of pent-up gases that were eventually vented through volcanoes. He recognized the connection, even though we would not agree today with his explanation. Perhaps the most important lesson of the past forty years has been the virtue of scientific humility: Who knows which ideas that now appear untenable will turn out to be right, and which of those ideas that we currently accept on the basis of available evidence will seem hopelessly naive 2000 years from now?

It is now recognized that most of the world's volcanoes occur along either the seams of converging plates (the *subduction zones*) or the seams of diverging plates (the *rift zones*). Hence, plate tectonics plays a major role in the location and nature of volcanic activity. The next five chapters amplify this tectonic framework. Chapter 2 describes the birth of Surtsey, a rift volcano in Iceland, and Chapter 3 explains the general features and processes related to rift volcanoes. Chapters 4 and 5 examine individual volcanoes—Krakatau and Mount St. Helens—near the converging edges of plates, and Chapter 6 looks at the general features and processes related to subduction volcanoes.

Sources

Glen, W. *The Road to Jaramillo*. Stanford, CA: Stanford University Press, 1982.

Gore, Rick. "Our Restless Planet Earth." *National Geographic Magazine*, August 1985.

Kearey, P., and F. J. Vine. *Global Tectonics*. Palo Alto, CA: Blackwell Scientific, 1990.

Kious, W. J., and R. I. Tilling. *This Dynamic Earth: The Story of Plate Tectonics*. Denver: U.S. Geological Survey, 1996.

Moores, E. M., ed. *Shaping the Earth: Tectonics of Continents and Oceans*. New York: Freeman, 1990.

Moores, E. M., and R. J. Twiss. *Tectonics*. New York: Freeman, 1995.

Plummer, C. C., and D. McGeary. *Physical Geology*, 7th ed. with CD-ROM. Dubuque, IA: Brown, 1996.

Simkin, T., J. D. Unger, R. I. Tilling, P. R. Vogt, and H. Spall. *This Dynamic Planet: World Map of Volcanoes, Earthquakes, Impact Craters, and Plate Tectonics*. Denver: U.S. Geological Survey, 1994.

Sullivan, W. *Continents in Motion*. New York: McGraw-Hill, 1991.

Surtsey, Iceland

10   Lightning bolts discharge the buildup of electricity in the uprushing ash column at Surtsey. Ninety-second time exposure. (Photograph by Sigurgeir Jónasson, December 1, 1963)

At three o'clock in the morning we saw smoke rising from the sea and thought
it to be land; but on closer consideration we concluded that this was a special
wonder wrought by God and that a natural sea could burn.
—Captain Jorgen Mindelberg, Iceland (1783)

Iceland sits astride the Mid-Atlantic Ridge, part of the worldwide rift system. This 300-by-500-kilometer island whose north shore touches the Arctic Circle is entirely volcanic in origin. Eruptions occur about every five years, sometimes from distinct volcanoes like Mount Hekla, which has erupted twenty-two times since A.D. 900, but often from long fissures that erupt only once. Iceland grows in size from volcanic eruptions and is worn away by the storms of the North Atlantic. Icelandic farmers and fishermen know that the Earth is a dynamic place; they all are geologists at heart.

The story of the Surtsey eruption begins at sea. The late Icelandic volcanologist Sigurdur Thorarinsson was its chronicler:

At 7:15 A.M. on November 14, 1963, Olafur Vestmann, the cook on a fishing boat off the south coast of Iceland, noticed dark smoke on the horizon against the half-light of dawn. Since there was no land in that direction, his first thought was that it was a ship on fire, but soon the fishermen recognized that the black columns of volcanic ash signaled an eruption from beneath the sea. By 11:00 A.M., the eruption cloud of ash and steam had reached a height of 3500 meters, and volcanologists were on their way to observe the eruption from the air.

At 11:30 A.M., the eruption was coming from three submarine vents along a northeast–southwest line. Explosions from the vents were occurring every few seconds, shooting jets of dense black ash to heights of 100 to 150 meters. By 3:00 P.M., the separate eruption columns had joined along a 500-meter line across the ocean's surface, and the rapid emission of black volcanic debris indicated that a new island was forming. Surtsey, named for Surtur, the mythological giant of fire, was born that night (Figure 11).

The eruption was a surprise. Only in hindsight were the few clues to the forthcoming eruption recognized. Two days before the eruption, a marine research vessel noticed a strange rise in the temperature of the sea surface above the normal 7° to 9°C, in an area about 3 kilometers from the eruption site. Also on November 12, people in Vik, a coastal village 80 kilometers to the east, had noticed the rotten-egg smell of hydrogen

11 Column of steam and ash rising more than 3000 meters from the Surtsey eruption on November 16, 1963, two days after the eruption was first sighted. (Photograph by Hjálmar Bárdarson)

sulfide. The seismograph at Reykjavik, 120 kilometers away, had recorded weak tremors a week before the eruption was first observed, but the location of the seismic disturbance could not be determined. No preliminary earthquakes were felt at the fishing port of Vestmannaeyjar on Heimaey Island, 22 kilometers northeast of Surtsey, the closest settlement to the eruption site (Figure 12).

Presumably the eruption began quietly at 130 meters below sea level and took days or weeks to build the volcano to just beneath the sea surface. There, the explosive activity could no longer be contained and quenched by the pressure and chill of the sea.

12   A. Map of the Vestmann Islands, showing Surtsey, which formed in 1963–1967, and Heimaey, which formed in prehistoric times. B. The shaded area represents lava added to Heimaey during the 1973 eruption. The former shore is shown by a dashed line.

During the first week, the new island erupted and grew almost continuously. Closely spaced explosions merging into a steady jetting of volcanic ash formed a towering column that rose to a height of 9 kilometers.

By November 18, Surtsey was a ridge 550 meters long and 45 meters high, split lengthwise by the erupting fissure. Gradually one vent along the fissure became dominant, and the island began to assume a more circular shape. On November 24, the island was 900 meters long and 650 meters wide, and the rim of the main crater was almost 100 meters high. The entire island was built of loose, dark volcanic debris, thrown out in fragments by the explosive eruptions. Lava flows of liquid rock had not yet begun, and the loose volcanic pile was an easy target for the strong winter storms of the North Atlantic. But the eruption more than kept pace with the marine erosion and slumping. By February 5, 1964, the maximum height of the volcanic island was 174 meters, and the maximum diameter was 1300 meters, including a wave-cut terrace about 150 meters wide around the shore.

During the first three months of the eruption, the sea had easy access to the erupting vents, by both direct flooding and seepage through the loose volcanic pile. This contact with water produced the billowing steam explosions that characterize shallow submarine eruptions. Each explosion expelled a black mass of rock fragments, out of which shot numerous larger fragments of pasty lava, called *volcanic bombs*. The bombs left trails of black volcanic ash that rapidly turned white and furry as the hot, invisible steam in the trails cooled and condensed. The sight of the black, jetlike arcs turning grayish white made the eruptions look like an exploding fireworks factory (Figure 13). The bombs whistled as they spun in flight, some landing in the sea more than a kilometer from the vent. The steam explosions themselves were peculiarly quiet, beginning with a muted thump and an almost silent burst of debris.

When the explosions occurred in rapid sequence, as they often did, a large cloud of ash and steam rose above the island. On the rare calm days this plume reached as high as 10 kilometers, but generally the strong winds bent the column at 500 to 2000 meters and caused ashfalls many kilometers downwind, contaminating the water supply caught on the roofs in Vestmannaeyjar. The violent updrafts of air into the ash cloud caused whirlwinds, waterspouts, lightning flashes, and sometimes hailstones of falling volcanic fragments coated with ice (Figure 10).

13   Black jets of ash explode to heights of 250 meters during the Surtsey eruption. Jet trails follow behind large volcanic bombs that whistle as they spin in flight. (Photograph by Thorleifur Einarsson)

At times the explosive eruptions changed into a continuous uprush of volcanic fragments and steam. Bombs, glowing red to orange, lit the rising columns at night. There were only brief lulls in the eruption during November and December, but by January, periods of repose lasted up to a day. As the island grew, the seawater was more effectively blocked from the crater, and the steam explosions diminished.

In April, fire fountains and lava flows became dominant, forming a cap of hard, solid rocks over the lower slopes of Surtsey (Figure 14). The island was now assured a place on the world's maps; the waves still pounded on the shores, but they met their match on the hardened lavas. The quiet effusion of lava continued for more than a year as

Surtsey increased in area to 2.5 square kilometers, more than half of this capped by hard lava flows (Figure 15).

During 1965 and 1966, small explosive eruptions built low islands of volcanic debris both northeast and southwest of Surtsey, but since these loose volcanic piles were not large or armored with lava flows, they were soon washed away by ocean storms.

In August 1966, lava again erupted from the main crater on Surtsey, and lava flowed continuously in tunnels beneath its own crust to the edges of the island. We visited Surtsey during this last activity and were amazed to see that the orange tongues of lava pouring from the tunnels into the sea still glowed as they plunged beneath the waves. The insulating effects of the lava's chilled skin and the steam layer on its surface were so effective that they prevented rapid quenching and steam explosions.

14   During 1964, Surtsey grew until the sea was blocked from the vent area. As a result, the eruption changed from explosions to incandescent lava fountains and flows. White steam clouds rising where the flows entered the sea replaced the black and gray explosion clouds. (Photograph by Captain Gardar Pálsson)

A

15   Aerial photographs looking down vertically on the island of Surtsey (about
1.3 kilometers wide, east to west, in photo A, north at top, August 25, 1964).
The gray area in photo A is volcanic ash and cinders, and the dark area in the
south is covered by basaltic lava flows from the crater. A second crater developed
to the east of the active vent (photo A) between 1964 and 1967. By July 21,
1996 (photo B), a large area of lava on the south side of the island had been
eroded away by the strong waves of the North Atlantic Ocean, and a sand spit
had grown on the lee side of Surtsey. (Photographs © by the National Land
Survey of Iceland)

B

By June 1967, the eruption was over; it had lasted for three and a half years. The total volume of volcanic ash and lava was slightly more than 1 cubic kilometer, only 9 percent of which was above sea level. The rates of eruption were highest in the early phases, diminishing more or less continuously during the course of the activity. The temperatures of the lava on emission averaged about 1140°C.

The eruption had three main phases. First, there was the quiet, undetected growth from 130 meters below sea level to just a few meters below sea level, possibly lasting for several weeks. Second, explosive eruptions from shallow water built an island of loose volcanic debris, the beginning of the observable eruption. Third, the blocking of water from the vent stopped the explosions and allowed the quiet emission of lava flows (Figure 16).

The shape of Surtsey reflects the island's three-part volcanic history. The sides of the underwater base are steep, owing to the rapid cooling and piling up of the submarine lavas. The explosive debris formed steep crater walls and cones above sea level but was cut into a flat bench at sea level by wave erosion. The final lava flows formed gentle slopes where they poured out and hardened over the loose debris from earlier explosions.

Surtsey was born in sea, steam, and fire, its rock ribs molded by the environment into which it erupted. This is also how the Earth is shaped by such conflicting forces, but over immense spans of time; it is rare to witness these events condensed into less than four years.

In one sense, the story of Surtsey is just beginning. The new island is closed to all but researchers studying the ways in which, even in this hostile environment, life arrives and slowly but tenaciously develops a complex ecological system. We recently asked our colleague Páll Einarsson, an Icelandic seismologist and volcanologist, about how life

16   (Facing page) Diagrammatic cross sections showing the three main phases of the eruption of Surtsey: A. The submarine volcano of pillow lava is just reaching the sea surface and becoming explosive. Pillow lava forms during submarine eruptions where the pressure of the overlying water prevents explosive boiling. B. Continuous jetting explosions from the contact of seawater with erupting lava excavate the crater downward into old seafloor sediment. C. The collapse of walls fills the deep crater, and the slumps are covered by surface flows of fluid basaltic lava. These cross sections are schematic profiles drawn along a west northwest–east southeast line through the center of the crater, as seen in Figure 15A and B. Data on the deposits that slumped into the deep crater (C) were obtained by drilling. (Adapted from J. G. Moore, Geological Magazine 122 [1985]: 655.)

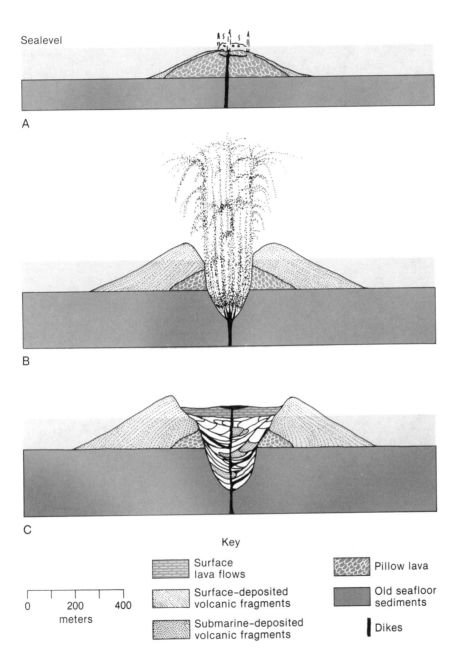

Sealevel

A

B

C

Key

Surface
lava flows

Surface-deposited
volcanic fragments

Submarine-deposited
volcanic fragments

Pillow lava

Old seafloor
sediments

Dikes

0   200   400
meters

has taken hold on Surtsey over the past thirty years. His answer was "slowly," and he elaborated as follows: Birds began nesting on the island three years after its formation, and the eight species now living there all depend on the sea for their food. Some soil has formed in the birds' small roosting places, and these are the only areas of continuous vegetative cover. Thirty species of higher plants, 70 mosses, 12 lichens, and 170 insects now call Surtsey home. Seals have been breeding on the island since 1983, but they are more transients than residents.

From empty sea, through chaos and barren island, to the slow beginnings of life, the story of Surtsey recapitulates in a small way the origin of the Earth.

Sources

Jakobsson, Sveinn, Sturla Fridriksson, and Erlingur Hauksson. *Surtsey 30 ára.* Reykjavík: Surtseyjarfélagid, 1993. (In Icelandic; soon to be translated into English.)

Thorarinsson, Sigurdur. *Surtsey.* New York: Viking, 1967.

Thorarinsson, Sigurdur. "Surtsey: Island Born of Fire." *National Geographic Magazine*, May 1965.

3

Fire Under the Sea

17 Hawaiian lava entering the sea. (Photograph by R. W. Grigg, University of Hawaii.)

In nature things move violently to their place, and calmly in their place.

—Francis Bacon (1561–1626)

W hat do Surtsey and other rift volcanoes tell us about the formation of the Earth? They tell us that the entire deep-ocean floor, 60 percent of the Earth's surface, is of volcanic origin—the product of rift volcanism.

The worldwide rift system is 60,000 kilometers long, and nearly all of it is submarine. Where it does appear on land, in places like Iceland, the volcanic activity is probably greater than it is along the submarine ridges, great enough in fact, to build the ridge above sea level. The reason for this difference in volcanic activity is a topic of lively debate, and we will return to it in Chapter 8. Most geologists now agree, however, that Iceland is reasonably typical of the volcanic and structural processes that occur on the mid-ocean ridges, and an island is certainly easier to study than an underwater ridge.

Geologists who have studied the ocean ridges believe that reservoirs of molten rock, known as a *magma chambers,* exist at depths as little as 1 to 3 kilometers beneath the ridge axes. The height and width of these shallow bodies of partly molten rock is only a few kilometers, but in total they are tens of thousands of kilometers long, separated into segments along the crest of the oceanic ridges. Molten rock from these chambers feeds into the spreading rift and temporarily scars over the cracks between the separating plates (Figure 18).

The number of volcanoes that erupt every year to fill these submarine cracks may be greater than that recorded on land. Although a deep submarine volcanic eruption has never been observed, a rough estimate of the number of such events can be made from the observation that Iceland averages one eruption every five years. Iceland represents about 1/200 of the length of the spreading oceanic ridges. To account for its extra elevation, assume that Iceland's eruption rate is twice that of an equal length of the submarine ridges, whose volcanoes therefore erupt once every 10 years. Thus 200 Icelands times 1/10 of an eruption per year equals 20 eruptions per year along the oceanic

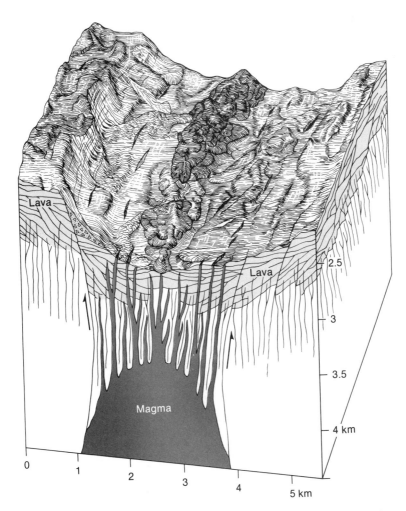

Lava

Lava

2.5

3

3.5

4 km

Magma

0   1   2   3   4   5 km

18   A diagram of the structure and submarine topography of the central rift valley of the Mid-Atlantic Ridge. The vertical scale is exaggerated two times. (After R. Hekinian, J. G. Moore, and W. B. Bryan, *Contributions to Mineralogy and Petrology* 58 [1976]: 107.)

ridge system. The fact that not one of these deep submarine eruptions has ever been directly witnessed demonstrates that many frontiers remain to be crossed in our exploration of the Earth. It is only a matter of time, however, until these deep submarine eruptions are seen and studied.

There already have been some close encounters. In 1993, sub-mariners exploring the crest of the East Pacific Rise at depths of 2500 meters below sea level and latitudes 17° to 19° south saw evidence of very recent eruptions of lava. In this area, the plates are separating at a rate of nearly 15 centimeters per year, and several very young appearing flows, completely devoid of any sediment cover, were observed. On the Juan de Fuca Ridge, off the coast of the northwestern United States, an earthquake swarm and thermal anomaly in June 1993 was interpreted to be a possible submarine lava eruption. Dives with submersibles have confirmed the eruption of a new 1.6-million-square-meter flow at that site.

The volume of volcanic rock needed to heal the tear between the spreading plates is enormous. Every year 2.5 square kilometers of new seafloor are formed by the spreading of the plates. The volcanic part of the rocky plates is 1 or 2 kilometers thick at their spreading edges; hence the volume of new volcanic rock amounts to about 4 cubic kilo-meters of new oceanic crust every year.

The description of the Surtsey eruption in Chapter 2 contains an interesting paradox. The eruption of molten rock from a vent at shallow depth in the sea led to violent steam explosions. But the lava that flowed through tunnels to the growing edge of Surtsey was seen to flow beneath the waves and to generate white steam clouds without violently disrupting the molten rock. Both were fire under the sea. Why was only the former explosive?

The answer involves the amount of water and other gases dissolved in the magma, about 1 percent of the weight of the molten rock in the Surtsey lavas. When this type of magma from depth is erupted into shal-low water, it quickly forms a chilled, glassy rind. At low surface pres-sures, most gases boil out of the molten rock and expand greatly in vol-ume. This expansion fractures the glassy rind of the venting magma and gives the seawater access to a large surface of finely broken but still hot rock. The steam generated by this contact water, added to the steam and other gases boiling out from the magma, creates the explosions.

Lava that has erupted from a crater on land and has then flowed through tubes to the sea has already lost most of its dissolved gases. As it flows into the sea, it forms sacklike bodies of lava with chilled, glassy rinds; these bodies are called *pillow lava* (Figure 19). These "pillows" do

A

B

19   Pillow lava formed by volcanic flows under water makes up some of the most widespread rocks on Earth. A. Pillow lava formed beneath the prehistoric glacial ice on Iceland. B. Young pillow lava on the Mid-Atlantic Ridge 2700 meters below the ocean surface. (Photograph taken from the submersible ALVIN by W. B. Bryan, Woods Hole Oceanographic Institution.)

not explode from the expansion of internal gases because those gases have already boiled away on land. A relatively small surface area loses heat to the surrounding water, and some steam is formed without explosions.

This process suggests that lavas emitted from vents in deep water—deep enough that the dissolved gases cannot boil and expand—also quietly form pillow lavas or sheetlike flows. Because the contact water is under too high pressure to boil, no steam forms. The violence of fire under the sea is thus controlled by depth. The commonsense notion that fire and water do not mix originates in the limited view of our surface pressure environment. Of course, the word *fire* is used loosely. In this context, fire means hot and glowing, not the combustion of a fuel with oxygen to produce flames. Deep under the ocean, fire and water find a way to exist side by side.

The critical depth below which submarine lava is not explosive depends on the dissolved gas content and the temperature of the magma, but for basalt lava with a low gas content, that depth is only about 30 meters below sea level. Most of the oceanic ridges where the seafloor is spreading and new oceanic crust is forming are more than 2000 meters below sea level. Pillow lavas and sheet flows thus form much of the pavement of the spreading seafloors. Rocks dredged from the ridges by oceanographic ships and sampled on recent dives by scientists in special submersibles confirm this theory (Figure 20).

Jim Moore, one of the divers who descended 3000 meters into the valley along the crest of the Mid-Atlantic Ridge, describes the scene as a strange, silent land of pillow lava hills and yawning cracks, with a dusting of sedimentary mud. The pillow lava forms steep-sided hills 20 to 30 meters high and one-half to 1 kilometer wide near the center of a 4-kilometer-wide rift valley. These hills appear fresh and recently erupted, clearly marking the very axis of the spreading oceanic ridge system.

On the valley floor on either side of the hills, many of the lavas look older. They are being covered with loose sedimentary debris that is slowly raining down from the ocean surface above. Most of the sediment is composed of microscopic shells of calcium carbonate from floating plants and animals. The relative age of the submarine lava flows can be established by the thickness of the sedimentary mud that slowly accumulates with time (Figure 21).

20   The submersible ALVIN in the rift valley of the Mid-Atlantic Ridge. This artist's conception shows ALVIN exploring an open fracture cutting across young pillow lava at a depth of 2600 meters. (Painting by the National Geographic Society.)

Deep cracks parallel to the rift valley cut across the older lavas. These cracks result from the pulling apart of the entire ocean-ridge system (Figures 22 and 23). The steep walls that form the sides of the rift valley appear to be normal faults along which the valley has dropped down about 500 meters relative to the ridge.

For more than 1000 kilometers on either side of the rift valley, the Mid-Atlantic Ridge slopes generally downward away from the ridge

21   View down into an open fracture in the rift valley of the Mid-Atlantic Ridge. The fracture cuts across sediment-covered pillow lava and was evidently formed by the forces separating the plates. This photo was taken at a depth of about 2700 meters during Project FAMOUS (French–American Mid-Ocean Undersea Study) in 1974. Similar above sea-level fractures cut through lava flows in the rift zones of Iceland. (Photograph by W. B. Bryan, Woods Hole Oceanographic Institution.)

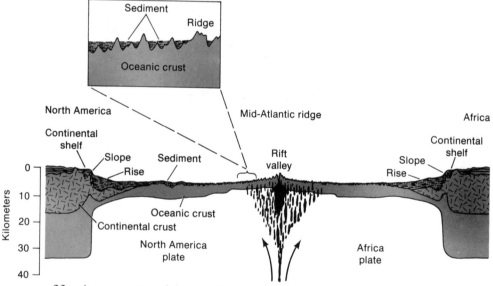

22   A cross section of the Earth beneath the Atlantic Ocean Basin. Both North America and Africa are moving with the spreading seafloor, forming "passive" continental margins. Sediments accumulating in the valleys between the volcanic ridges slowly bury the rugged submarine mountains. (After B. C. Heezen, in F. Press and R. Siever, *Earth*, 2nd ed., p. 262. W. H. Freeman and Company, Copyright © 1978.)

Axial
rift

Sediments

Pillow lavas and
sheet flows

Sheeted dikes

Normal fault

Gabbro

Magma Chamber

Thickness (kilometers)

.25

1.5

.5

5

23 A schematic cross section of a spreading mid-ocean ridge. Active submarine volcanism is concentrated in a narrow area along the rift axis. Faulting forms the down-dropped blocks of older volcanic rock on both sides of the spreading axis. The magma forms as the plates diverge and mantle rock rises and melts with the decrease of pressure. It collects in a chamber below the spreading center. Along the borders of the magma chamber, a coarse-crystalline rock of basaltic composition called *gabbro* is formed. At the top of the chamber, the magma rises as the plates diverge and cools as vertical dikes. At the surface, lava flows out and hardens in the form of sheets and pillows. As the new crust moves away from the spreading center, a layer of sediments is deposited on it. The mature crust has a layered structure from the top down: sediments, sheet and pillow lavas, dikes, and gabbros. (From Jean Francheteau, "The Oceanic Crust. East Pacific Rise." Copyright © 1983 by Scientific American, Inc. All rights reserved.)

over rugged and fractured volcanic terrain into deep basins 4000 to 5000 meters below sea level.

The East Pacific Rise spreads at a faster rate than does the Mid-Atlantic Ridge; its magnetic stripes indicate separation rates of 6 to 17 centimeters per year. Recent exploration of this rift system has revealed some interesting similarities and differences between Atlantic and Pacific seafloor spreading (Figure 24).

Exploratory dives in the submersible named ALVIN, operated by the Woods Hole Oceanographic Institution, were concentrated in an area just south of Baja California, at a depth of 2500 meters. There the rift axis is only about 1 kilometer wide, and the spreading of the East Pacific Rise is 6 centimeters per year—about twice the spreading rate of the Mid-Atlantic Ridge.

A
B

24   Comparison of the topography of the axial areas of a fast-spreading mid-ocean ridge (A, the East Pacific Rise) to a slow-spreading ridge (B, the Mid-Atlantic Ridge). The axis of a fast-spreading ridge is an elevated zone possibly supported by the buoyant force of a shallow magma chamber beneath the axis. This axis is often offset 50 kilometers or more by major transform faults (A, bottom). Smaller offsets (A, top) are formed by overlapping ridge segments. The axis of a slow-spreading ridge is a valley. It is also offset by transform faults (B, bottom) or bends along the axis (B, top). (From Ken Macdonald and Paul Fox, "The Mid-Ocean Ridge," *Scientific American,* June 1990.)

Besides young pillow lava, the divers found some smooth-surfaced basaltic lava flows and solidified lava lakes. A fracture zone with fissures generally parallel to the rift system extends one-half to 2 kilometers on each side of the youngest lavas. Major fault scarps, some as high as 70 meters and facing in toward the rift axis, extend outward from the spreading center. The locations of earthquakes and sea-bottom evidence of active faulting indicate that the spreading is concentrated within a 20-kilometer-wide zone, mostly near the central axis. Beyond this zone, the plates continue to move passively away on their long journey beneath the Pacific Ocean.

Submarine hot springs were also found on this expedition. Exotic life-forms and metal sulfide deposits associated with these hydrothermal vents are among the most exciting scientific discoveries of the past two decades. Warm springs (20°C) were first observed during dives on the Galapagos Ridge in 1977, and in 1979 extremely hot springs

(350°C) were observed during dives at 21° north latitude on the East Pacific Rise. Bruce Luyendyk of the University of California at Santa Barbara was one of the scientists on the discovery dives, and his own words best describe the experience:

> The scene was like one out of an old horror movie. Shimmering water rose between the basaltic pillows along the axis of the neovolcanic zone. Large white clams as much as 30 centimeters long nestled between the black pillows, while crabs scampered blindly across the volcanic terrain. Most dramatic of all were the clusters of giant tube worms, some of them as long as 3 meters.

On a subsequent dive, the scientific submariners found another impressive sight: "extremely hot fluids, blackened by sulfide precipitates, were blasting upward through chimneylike vents as much as 10 meters tall and 40 centimeters wide. We named the vents 'black smokers'" (Figure 25 and Color Plate 25).

In the years since, several other undersea expeditions have discovered these strange oases of life and sulfide metal deposits at scattered places along the axes of spreading rifts in both the Atlantic and Pacific Oceans. They are apparently common but not continuous features of the entire world rift system and are most likely to be found where recent and frequent volcanic activity provide the heat and gases needed to sustain their exotic food chain and mineral deposition.

The significance of these strange colonies of life on the dark seafloor and their associated mineral deposits is just beginning to be appreciated fully by biologists and geologists. The bacteria that feed the food chain of strange worms, clams, and crabs live on volcanic heat and gases; they do not need sunlight (Figure 26). Before this discovery, the sun and photosynthesis were considered to be the essential suppliers of energy to life on Earth.

The very latest twist in this fantastic discovery is proof that some of the bacteria that sustain this food chain are not bacteria at all but a previously unknown form of life. This puzzling one-celled organism, named Archaea, was found to have genes distinct from those of any other known forms of life. Craig Venter—of the Institute of Genomic Research and leader of the team of scientists that mapped the 1738 genes in Archaea that were cultured from submarine vent samples—remarked that "two thirds of its genes don't look like anything we've ever seen in biology

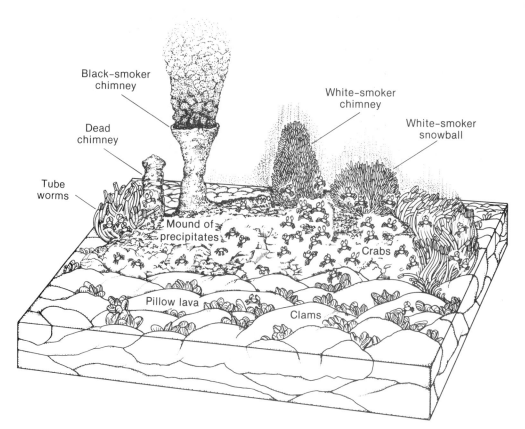

25   Idealized scene of living creatures and metal sulfide precipitates at a submarine hot spring 2500 meters deep along the rift zone of the East Pacific Rise. The white-smoker chimney is made of burrows of Pompeii worms. White, cloudy fluids emitted by the white smokers have temperatures up to 300°C. The hottest water, up to 350°C, pours from the black smoker and precipitates a dark cloud of metal sulfide particles as it is cooled by the surrounding seawater. (From K. C. Macdonald and B. P. Luyendyk, "The Crest of the East Pacific Rise." Copyright © 1981 by Scientific American, Inc. All rights reserved.)

before." Some of the genes of the more familiar third are similar to those in humans, and others are similar to those in bacteria. The Archaea of the submarine oases thrive in total darkness on a diet of carbon dioxide, hydrogen, nitrogen, and volcanic heat. The main by-product of their metabolism is methane.

26 Tube worms and other unusual life-forms flourish near submarine hot springs. This photograph shows clusters of large tube worms, some as long as 2 meters, waving in warm water at the margins of high-temperature vents. Chemosynthetic bacteria form the base of this sunless food chain that also supports clams and crabs. (Photograph taken at 21° north latitude on the East Pacific Rise during the RISE Expedition, F. N. Speiss, Scripps Institution of Oceanography, 1979, and provided by William Normark, U.S. Geological Survey.)

The metallic sulfide deposits are an equally important discovery. They precipitate where seawater chills the submarine hot springs, forming mounds and crusts that are rich in iron, copper, and zinc. As the seafloor separates, these deposits are spread into extensive layers of potentially great economic value (more on this in Chapter 15). One of the important lessons from the submarine exploration of the mid-ocean ridges is the value of serendipity—while searching for one thing, other things are discovered, which sometimes may be even more important than what was originally sought.

These recent submarine explorations and new detailed maps of the seafloor provide a good picture of the topography and structure of the mid-ocean ridges. In the faster-spreading ridges, like the East Pacific Rise, that diverge from 6 to 17 centimeters per year, the axis of the system is a ridge several hundred meters high and 5 to 20 kilometers wide.

This spreading axis is interrupted every few hundred kilometers by transform faults that offset the axis by many tens of kilometers. Within the segments separated by transform faults are smaller undulations and offsets of the central ridge crest, about every 100 kilometers. These may be related to segmentation of the magma supply reservoirs (Figure 27).

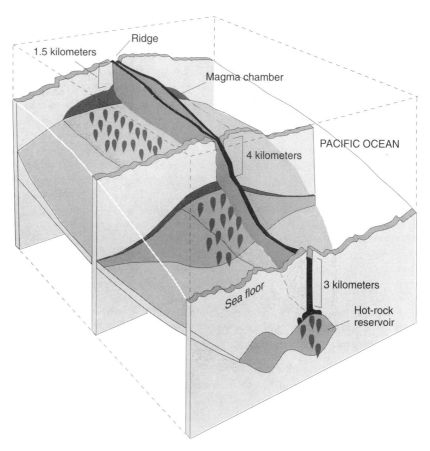

27   A schematic view of a magma chamber thought to occur in elongated segments beneath the axis of a fast-spreading mid-ocean ridge. The lens of magma at the top of the partly molten, hot-rock reservoir is fed by rising bodies of basalt. The magma lens is about 1.5 kilometers beneath the seafloor along the middle of a segment (far end of diagram) and about 3 kilometers below the seafloor near the end of a segment. The axial segments are 10 to 200 kilometers long and are offset 2 to 3 kilometers at their overlapping ends (From Ken Macdonald and Paul Fox, "The Mid-Ocean Ridge," *Scientific American,* June 1990.)

In the slower-spreading ridges, like the Mid-Atlantic Ridge, that diverge at about 3 centimeters per year, the axis of the system is a rift valley a few kilometers deep and 30 to 40 kilometers wide. The depth to the magma reservoirs is apparently greater than in the fast-spreading ridges, and these chambers seem to be spaced about every hundred kilometers. Both slow- and fast-spreading ridges are offset by major transform faults. Overall, the great mid-ocean ridge system is like the seam of a baseball. It is a nearly continuous feature, broken only into names as it extends through the ocean basins of the world.

One of the main arguments concerning rift volcanoes centers on whether they are a cause or an effect of the rifting apart of the oceanic ridges. Does the intrusion of volcanic rocks push apart the plates, or does the pulling or dragging apart of the plates form cracks through which the magma can rise to the surface? Most experts now favor the latter view, although the answer may be both.

During the past thirty years, Icelandic scientists have been measuring the spreading in their country, using sensitive surveying instruments that can detect changes of about 1 centimeter over distances of many kilometers. The results of these studies indicate that the rift zones slowly widen and sink a few millimeters per year across distances of hundreds of kilometers, without earthquakes or volcanic eruptions. When the tension reaches the breaking point, the rift fractures (Figure 28). A rifting event like this takes place in a few hours or days. Open cracks form parallel to the rift; earthquake swarms jolt the local area; and lava erupts from some of the fissures.

We were lucky enough to be in Iceland when one of these rifting episodes occurred on September 7, 1977. An earthquake swarm at Krafla Volcano was followed in rapid succession by a brief but spectacular volcanic fissure eruption with the ground cracking parallel to the rift (Figure 29). In a zone about 2 kilometers across and 20 kilometers long, the cracking formed open fractures as wide as 20 centimeters with vertical offsets as high as 1 meter. All this occurred between 3 P.M. and midnight.

One of the cracks cut through a producing geothermal steam field. Some magma was injected deep along this fracture, and one of the steam wells erupted small bursts of incandescent lava. The erupting lava cut through the heavy metal pipe at the wellhead and showered a few tons of frothy lava fragments like black popcorn around the well.

28   Aerial photograph of fractures in the rift zone of Iceland. These fractures mark the continuation of the crest of the Mid-Atlantic Ridge across Iceland. The lava flows in this area are less than 6000 years old. Summer homes dot the edge of the lake formed by the central valley of the rift.

This is the only known case in which lava has erupted from a man-made vent (Figure 30).

A survey line about 4 kilometers long that crossed the fracture zone lengthened by more than 1 meter, but survey lines on the flanks of the new fracture zone showed contraction. The picture of rifting that emerges is one of slow stretching that takes place over a wide zone for hundreds of years, followed by rupture that concentrates the deformation into a narrow central zone and allows the flanks to contract back to an unstretched condition. It is like pulling a rubber band slowly and then cutting it in the middle; all the stretching that had slowly accumulated is suddenly concentrated in the cut, and the sides contract. Perhaps a sudden tear in a piece of elastic cloth when it is slowly stretched would be a better analogy, since its dimensions fit the plates more closely than does a rubber band.

The length of the tear in a rifting episode is tens of kilometers long, only a very small fraction of the length of the entire rift system. The

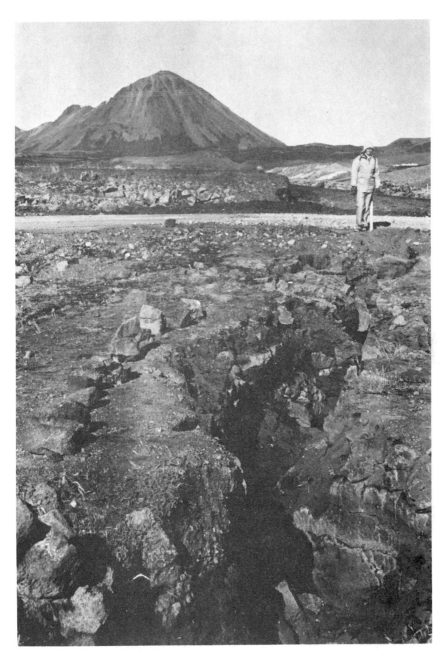

29 A gaping fracture formed during an earthquake swarm in northern Iceland on September 7, 1977. This rifting event was related to an eruption of Krafla Volcano. The cone in the background was formed during a prehistoric eruption of Krafla.

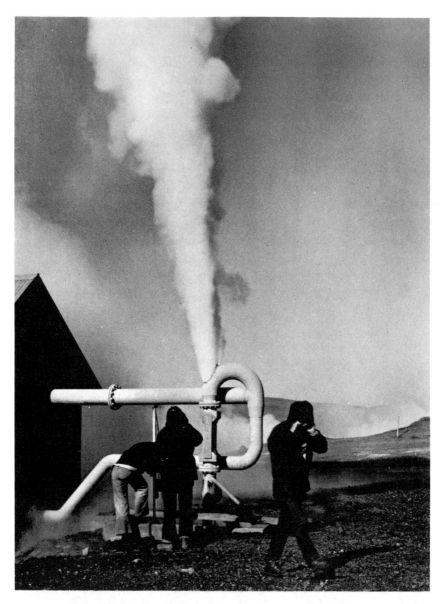

30   Incandescent lava was briefly ejected from this geothermal steam well in Iceland during the eruption of Krafla Volcano on September 7, 1977. A fracture somewhere between the surface and the 1200-meter depth of the well broke the well casing and allowed a small amount of molten rock to enter the producing well. The steam lifted the magma spray to the surface where it melted and abraded its way through the pipe. The few tons of small lava bombs that were ejected form the granular surface material. This is the only known case of a volcanic vent produced by drilling.

rifting episodes are infrequent in any one place, perhaps occurring only once every few hundred years. In any one year, however, rifting probably takes place somewhere along the length of the ocean ridge system.

The presence of magma under pressure beneath the rift zones probably aids the rupturing process, just as a wedge speeds the felling of a leaning tree. The processes of rifting and volcanism are so interrelated that it is difficult to sort out cause and effect—reminding us of Samuel Butler's remark that "a hen is just an egg's way of making another egg."

There is no doubt, however, that great chains of volcanoes, mostly submarine, circle the Earth along the crests of the oceanic ridges. They owe their origin to the still mysterious processes that spread the plates. Surtsey is only a visible tip of this vast hidden activity.

Sources

Cone, Joseph. *Fire Under the Sea*. New York: Morrow, 1991.

Macdonald, Ken C., and Bruce P. Luyendyk. "The Crest of the East Pacific Rise." *Scientific American*, May 1981.

Macdonald, Ken C., and Paul Fox. "The Mid-Ocean Ridge." *Scientific American*, June 1990.

Nicolas, Adolphe. *The Mid-Ocean Ridges*. New York: Springer-Verlag, 1994.

Solomon, S. C., and D. Toomey. *The Structure of the Mid-Ocean Ridges*. Palo Alto, CA: Annual Reviews of Earth and Planetary Science, 1992.

4

Krakatau, Indonesia

31 Anak Krakatau in 1960.

Had the fierce ashes of some fiery peak
Been hurl'd so high they ranged about the globe?
For day by day, thro' many a blood-red eve,
The wrathful sunset glared. . . .

—Tennyson (1809–1892)

During the fall of 1883 and much of 1884, sunsets around the world were so spectacular that artists in Europe were inspired to paint their lurid glow. The cause of these unusual sunsets was a topic of lively debate, and even the British poet Alfred Tennyson pondered their origin in verse. Was it an island volcano half a world away that the British called Krakatoa?

In less than a day, Krakatau (its Indonesian name) disgorged 18 cubic kilometers of rock, produced waves more than 30 meters high, and formed a circular caldera 6 kilometers in diameter and 1 kilometer deep. In that short time, Krakatau erupted more volcanic rock than is formed along all the oceanic rifts in one year. Even though the Earth evolves slowly, much of the change is the sum of many brief, catastrophic moments.

The explosion of Krakatau in 1883 produced worldwide effects. The noise was heard for thousands of kilometers, and the shock wave was recorded on barographs around the world. The dust that was lifted into the stratosphere circled the globe, producing astonishing visual effects—colorful sunrises and sunsets and a curious, blue green appearance to the sun and moon. This dust spread westward, encircling the equator in two weeks, and then drifted both north and south. The average solar radiation in Europe decreased 10 percent over the next three years, and the average world temperatures were below normal. For the first time, scientists began to accept the idea that a volcanic eruption could have worldwide effects.

Krakatau, a small group of uninhabited volcanic islands in the Sunda Straits between Java and Sumatra, lies on a main oceanic trade route between Europe and Asia. The highest peak on the islands was less than 1000 meters tall—not an imposing cone—and there had not been an eruption there since 1680.

On May 20, 1883, Krakatau began to erupt with small explosions of ash from a low crater on the north end of the main island. Dutch scientists visited the island on May 27 and noted that much of the vegetation had been killed, but not burned, by the accumulation of fine volcanic ash. They also noticed some blocks of pumice, a rock composed of glassy froth so filled with bubble holes that it floats on water.

The activity declined until late June, when passing ships reported two columns of steam. Apparently a new vent had also become active. On August 11, when the island was again visited, three vents were producing small explosive eruptions. Not much additional ash had accumulated since May, and some trees were still alive on the southern peak of the island.

Things changed dramatically on August 26. At 1 P.M. noises like thunder were heard up to 200 kilometers away, and by 2 P.M. a black cloud had climbed to 27 kilometers above Krakatau. The closest witnesses were aboard the British ship *Charles Bal*, bound for Hong Kong (Figure 32). The captain's log gives some vivid impressions of the eruption's climax:

(August 26) At noon, wind west–southwest, weather fine, the Island of Krakatoa to the northeast of us, but only a small portion of the northeast point, close to the water, showing; rest of the island covered with a dense black cloud. At 2:30 P.M. noticed some agitation about the Point of Krakatoa; clouds or something being propelled with amazing velocity to the northeast. To us it looked like blinding rain, and had the appearance of a furious squall of ashen hue. At once shortened sail to topsails and foresail. At 5 P.M. the roaring noise continued and increased, darkness spread over the sky, and a hail of pumice stone fell on us, many pieces being of considerable size and quite warm. Had to cover up the skylights to save the glass, while feet and head had to be protected with boots and southwesters. About 6 P.M. the fall of larger stones ceased, but there continued a steady fall of a smaller kind, most blinding to the eyes, and covering the decks with 3 to 4 inches very speedily, while an intense blackness covered the sky and sea. Sailed on our course until we got what we thought was a sight of Fourth Point Light; then brought the ship to the wind, southwest, as we could not see any distance, and we knew not what might be in the Straits, the night being a fearful one. The blinding fall of sand and stones, the intense blackness above and around us, broken only by the incessant glare of varied kinds of lightning and the continued explosive roar of Krakatoa, made our situation a truly awful one. At 11 P.M., having stood off from the

32   The 1883 eruption of Krakatau generated giant sea waves that killed 36,000 people. The shaded areas on the coasts of Java and Sumatra were the zones inundated by tsunamis from the great explosive eruptions and caldera collapse. (After C. J. Symons, ed., *The Eruption of Krakatoa*, Royal Society Report of the Krakatoa Committee, 1888.)

Java shore, wind strong from the southwest, the island, eleven miles west–northwest, became more visible, chains of fire appearing to ascend and descend between the sky and it, while on the southwest end there seemed to be a continued roll of balls of white fire; the wind, though strong, was hot and choking, sulphurous, with a smell like burning cinders. From midnight to 4 A.M. the same impenetrable darkness continuing, the roaring of Krakatoa less continuous but more explosive in sound, the sky one second intense blackness and the next a blaze of fire; mastheads and yardarms studded with electrical glows and a peculiar pinky flame coming from clouds which seemed to touch the mastheads and yardarms. At 6 A.M., being able to make out the Java shore, set sail. Passed Anjer at 8:30 A.M., close enough in to make out the houses, but could see no movement of any kind. At 11:15 there was a fearful explosion in the direction of Krakatoa, now over 30 miles distant. We saw a wave rush right on to Button Island, apparently sweeping right over the south part and rising half way up the north and east sides. This we saw repeated twice, but the helmsman says he saw it once before we looked. The same waves seemed also to run right on to the Java shore. The sky rapidly covered in, by 11:30 A.M. we were inclosed in a darkness that might almost be felt. At the same time commenced a downpour of mud, sand, and I know not what; ship going northeast by north, seven knots under three lower topsails; put out the side lights, placed two men on the lookout forward, while mate and second mate looked out on either quarter, and one man employed washing the mud off binnacle glass. We had seen two vessels to the north and northwest of us before the sky closed in, adding much to the anxiety of our position. At noon the darkness was so intense that we had to grope about the decks, and although speaking to each other on the poop, yet could not see each other. This horrible state and downpour of mud continued until 1:30 P.M., the roarings of the volcano and lightnings being something fearful. By 2 P.M. we could see the yards aloft, and the fall of mud ceased. Up to midnight the sky hung dark and heavy, a little sand falling at times, the roaring of the volcano very distinct although we were fully sixty-five or seventy miles northeast from it. Such darkness and time of it few would conceive, and many, I dare say, would disbelieve, The ship, from truck to waterline, is as if cemented; spars, sails, blocks, and ropes in a terrible mess; but, thank God, nobody hurt or ship damaged. On the other hand, how fares it with Anjer, Merak, and the other little villages on the Java coast? (pp. 140–141)

Not well. More than 36,000 people had been drowned by the great ocean waves that were generated by the major explosion and collapse. This huge tsunami was the real catastrophe of Krakatau. Ships in the area were not swamped, but slowly rode up and down with the swells, whose crests and troughs were kilometers apart. But as these waves reached shallow water, particularly in the bays that faced Krakatau, they began to curl and break, forming huge surf waves that swept inland and washed over everything in their reach. In some locations where they were channeled into narrowing bays, they ran up to elevations more than 30 meters above sea level.

The 10 A.M. explosion was the largest natural concussion ever recorded. The ash cloud rose 60 kilometers above Krakatau, and the detonation was heard as far away as Australia. The main tsunami swept the coasts of Java and Sumatra about half an hour after the explosion (Figure 33).

At 10:50 A.M., there was a second huge explosion, but apparently this one did not generate a tsunami. Throughout the afternoon and night of August 27, explosions of diminishing intensity recurred and finally ceased. A few small eruptions continued into September and October, but most of the violence was packed into the time between 1 P.M. on August 26 and 11 A.M. on August 27.

The Sunda Straits were choked with floating pumice. As ships began to make their way through them, sailors soon noticed that most of Krakatau was missing. Where once the peak of Danan had reached 450 meters high, there was ocean 200 meters deep. A 5-by-8-kilometer

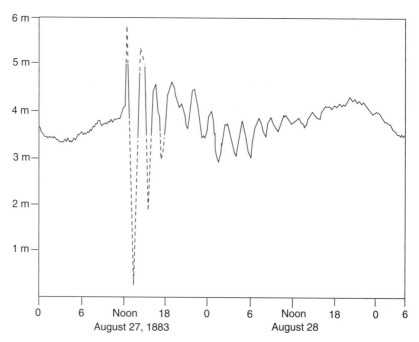

33 Tsunamis generated by the eruption of Krakatau recorded on the tide gauge at Djakarta (Batavia in 1883), 170 kilometers from the source, August 27–29, 1883. The broken lines are off scale but were reconstructed from high-water marks and eyewitness accounts. (After C. J. Symons, ed., *The Eruption of Krakatoa*, Royal Society Report of the Krakatoa Committee, 1888.)

chunk of the main island had disappeared, and a curving cliff, 800 meters high, cut through the south peak down to sea level. The origin of this caldera has been a major subject of debate ever since the 1883 eruption (Figure 34).

Some people thought that the island had simply blown its top and the debris had scattered into blocks and ashes. More than 90 percent of the erupted material, however, was pumice of a texture and composition not found in the wreck of the old volcano. It must have come, therefore, from the magma chamber beneath the volcano: when the volcano had emptied this chamber, the top collapsed into the void below. The great tsunami was probably generated by the collapse that followed the largest explosion. Most geologists now believe that calderas are largely formed by collapse. These great circular basins, some as large as 20 kilometers in diameter, are common volcanic features and attest to even greater eruptions than Krakatau's that took place in prehistoric times.

The Krakatau pumice covered thousands of square kilometers and was calculated to have a volume of 18 cubic kilometers. The volume of island that disappeared was about 6 cubic kilometers, a difference that can be explained by the fact that expanded pumice has three times the volume of its solid rock equivalent.

Some students of Krakatau have speculated that seawater was important in generating the explosions, as was the case at Surtsey. But the expansion of the pumice into a frothy glass is so complete throughout the scattered blocks that the expanding gases must have come from within the erupting rock. Rather than forming steam from contact with seawater, a high content of water dissolved in the magma appears instead to have propelled the explosions. Crater Lake in Oregon and many other calderas were formed by explosion and collapse in locations well inland, far from the sea.

On the remaining islands of Krakatau that surround the caldera, the deposits of explosive debris are tens to hundreds of meters thick. The lower layers are distinct and composed of well-sorted pumice fragments, indicating fallout from an eruption cloud. Each explosion added a separate and distinct layer. The debris from the two final huge explosions, however, is poorly sorted and only vaguely layered, because the pumice and fine ash were produced so rapidly that little sorting took place. This debris was deposited by great pyroclastic flows of expanding gases loaded with both fine and coarse volcanic fragments all mixed

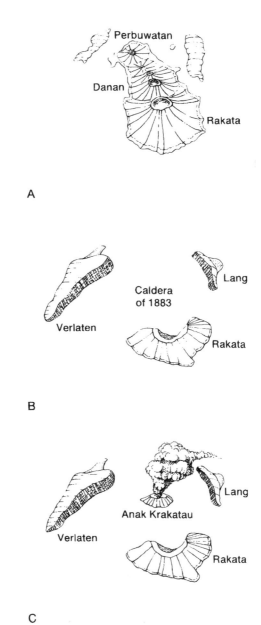

34 Changes at Krakatau since 1880. A. Krakatau before 1883. B. Krakatau after the 1883 eruption. C. Krakatau today. Before the eruption, Rakata Peak was about 800 meters high, and the main island about 9 kilometers long. Anak Krakatau, the currently active island, began its formation with submarine eruptions in 1927.

together. There were no lava flows. The rapid expansion of gas in the magma forms the pumice lumps by cooling the magma to semisolid but still hot glass fragments. The pumice, fine ash, and escaping gases produce a heavier-than-air, fluidlike emulsion, much like a snow avalanche, except that it is fiercely hot. These fast-moving pyroclastic flows are composed of newly erupted volcanic debris and are the most awesome and destructive of all volcanic phenomena. Huge clouds of finer ash fragments are lifted high above the pyroclastic flows by the heat and turbulence.

Australians 4800 kilometers away heard the largest Krakatau explosion as the sound of cannon firing. The shock wave rattled doors and broke windows in west Java and was recorded on barographs around the Earth. The more sensitive instruments recorded two or three world-circling trips of the airwave, each circuit taking about 36 hours. A 58-megaton nuclear test in the atmosphere over northern Russia in 1961 produced similar world-circling air-pressure waves.

Writers often compare the powers of nature with man-made explosions, but this is very misleading. The thermal energy in the 18 cubic kilometers of hot material erupted by Krakatau is equal to that of nearly 5000 megatons of hydrogen bombs. However, only a small fraction of Krakatau's thermal energy was converted to mechanical energy, and its release was spread out over a day, in contrast to the almost instant release of energy in a nuclear bomb. Therefore, the power (energy divided by time) of a nuclear bomb is much greater than the power of the 1883 Krakatau eruption. Even worse, the bomb creates radioactive debris.

Quiet prevailed at Krakatau for forty-four years; then in 1927 small explosive submarine eruptions began on the north rim of the caldera. Soon a small cinder cone was built above sea level and was named Anak Krakatau—"child of Krakatau." This small island has had a fitful history over the past seventy years, sometimes growing during periods of small ash and cinder explosions and sometimes being nearly washed away by waves from the Indian Ocean (Figures 35 and 36).

In 1960 we visited the new island during an eruptive episode. Small explosions of fragments ranging from fine ash up to 2-meter blocks were occurring at 30-second to 10-minute intervals. These explosions formed turbulent ash clouds that rose from 150 to 1200 meters above the 1.5-kilometer-diameter island (Figures 31 and 37). Clearly, Krakatau

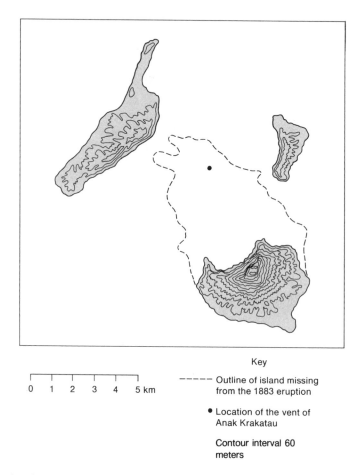

Key

0   1   2   3   4   5 km

----- Outline of island missing
from the 1883 eruption

● Location of the vent of
Anak Krakatau

Contour interval 60
meters

35   Sketch map of the Krakatau Islands after the great 1883 eruption. The missing area is 23 square kilometers; where a volcanic peak 450 meters high once stood, the water is now 275 meters deep. Anak Krakatau—"child of Krakatau"—began erupting in 1927. An island appeared in 1930 and has grown by means of small explosive eruptions and lava flows to a height of 200 meters and a diameter of nearly 2 kilometers.

is in a rebuilding phase, but it is in no rush compared with 1883. The volume of the new island is only about 1 cubic kilometer.

Since 1960, lava flows have issued from the crater and formed a protective cap on part of the island. In fact, the growth of Anak Krakatau has been similar to that of Surtsey. The lavas retard the wave ero-

36 Aerial view of Anak Krakatau in June 1959. Rakata Island and the 1883 caldera wall are in the background.

sion and give the new Krakatau a slightly more permanent hold on a notoriously impermanent spot in the world. Anak Krakatau has even become a tourist attraction, except during part of 1995 when Indonesian authorities closed the island to visitors because of renewed eruptions. During May 1995, 7200 explosions were recorded that shot columns of ash and volcanic bombs (large blobs of molten rock) more than 200 meters above the vent, and in September 1996 an airplane pilot reported a 3.7-kilometer-high eruption cloud from Anak Krakatau.

Will Krakatau ever erupt again as violently as it did in 1883? Probably not for thousands of years. The current composition of the lava indicates that most of the highly explosive magma was expelled in 1883, and new magma of this type forms slowly. Other volcanoes on the Ring of Fire may be approaching the onset of huge explosive eruptions, however. Calderas are a common volcanic feature, and a new one can be expected to form every few centuries.

37 Moderate explosive eruption of Anak Krakatau in January 1960, forming a 1000-meter-high ash cloud. The energy released in this explosion was less than one-millionth of the energy of the 1883 eruption.

Sources

Furneaux, R. *Krakatoa*. Englewood Cliffs, NJ: Prentice-Hall, 1964.

Simkin, Tom, and R. S. Fiske. *Krakatau 1883—The Volcanic Eruption and Its Effects*. Washington, DC: Smithsonian Institution Press, 1983.

Symons, C. J., ed. *The Eruption of Krakatoa*. London: Royal Society Report of the Krakatoa Committee. 1888.

Tjetjep, W. S. "Krakatau." *Bulletin of Global Volcanism Network* 21 (1996).

5

Mount St. Helens

38   On May 18, 1980, a giant explosive eruption and avalanche destroyed in moments the entire region surrounding Spirit Lake below Mount St. Helens. Not a tree was left standing; devastation was everywhere. But Earth abides; life will return to Spirit Lake over the centuries, as it did before. (Photograph by the U.S. Geological Survey.)

Men argue; Nature acts.
—Voltaire (1694–1778)

The violent eruption of Washington State's Mount St. Helens in May 1980 was one of the world's most closely monitored—and most fully documented—volcanic events (Figures 39 and 40). Volcanologists had been watching the mountain with keen interest, alerted by two months of mild but ominous activity that signaled the awakening of the sleeping giant (Figure 41).

At 8:32 A.M. on May 18, Mount St. Helens was shaken by a magnitude 5 earthquake, centered 1 to 2 kilometers beneath its north flank. Geologists Keith and Dorothy Stoffel, who were flying in a light plane about 400 meters above the summit at just that moment, saw several small icefalls start down the steep crater walls. About fifteen seconds later they were the closest witnesses to the onset of the largest landslide in recorded history, closely followed by a huge volcanic eruption (Figure 40).

"The whole north side of the summit crater began to move instantaneously as one gigantic mass. The nature of movement was eerie; not like anything we'd ever seen before," Dorothy recalled. "The entire mass began to ripple and churn without moving laterally. Then the whole north side of the summit started moving to the north along a deep-seated slide plane" (Figure 42B).

Seconds later, a massive explosion shook the mountain. From the Stoffels' viewpoint, the initial explosion cloud seemed to mushroom sideways to the north and plunge down the slope, but in the plane the Stoffels neither felt nor heard the explosion (Figure 42C). Realizing the enormous size of the eruption, survival became their first concern. They dove at full throttle to gain speed, but the expanding cloud appeared to be gaining on them; by turning south they finally outran it. Behind them, the rapidly growing ash clouds thrust north and northwest. To the east, the black clouds rose into billowing mushroom shapes, and lightning bolts thousands of meters high shot through them. Half

39 Map of the northwest United States showing the Cascade Range volcanoes and their relationship to the subduction of the Juan de Fuca Plate beneath the North American Plate. Submarine volcanoes and hot springs have recently been discovered along the rift zone that forms the west edge of the Juan de Fuca Plate. The dark gray areas are volcanic rocks less than 2 million years old. (Data from the U.S. Geological Survey.)

an hour later, the Stoffels, shaken but safe, landed at the Portland, Oregon, airport.

In the avalanche they had witnessed from the air, nearly 3 cubic kilometers of crushed rock and glacier ice plunged down into Spirit Lake and the North Fork of the Toutle River. Exploding steam helped

40   *(Facing page)* This remarkable sequence of photographs of the explosion of Mount St. Helens was taken by Keith Ronnholm, a geophysicist from Seattle, Washington, on the morning of May 18, 1980. He was camped near Bear Meadows, 18 kilometers northeast of the peak. After taking the pictures, Ronnholm was enveloped in the blackout of heavy ash and pumice fall, but he was just beyond the area of total devastation. The top photograph, taken 40 seconds after the great avalanche began, shows the huge landslide blocks being overtopped by the subsequent explosion. The bottom photograph, taken 20 seconds later, shows the rapidly growing explosion cloud thrusting northward from Mount St. Helens. (Photographs by Keith Ronnholm, © 1980.)

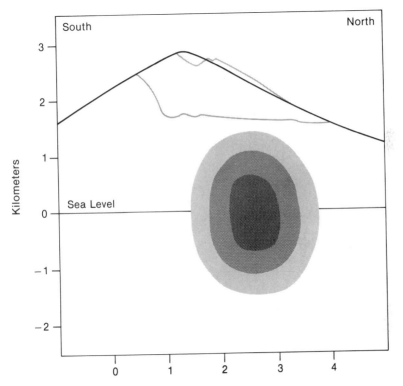

41   This cross section of Mount St. Helens shows the changes associated with the great 1980 eruption. The solid line at the top indicates the surface as of 1979. The upper broken line shows the crater and bulge that developed between late March and early May 1980. The shaded area indicates the region where the thousands of earthquakes that accompanied the growing bulge occurred. The darker the shading, the higher the density of earthquake locations. The lower broken line shows the crater after the May 18 eruption. (Data from the U.S. Geological Survey and University of Washington Department of Geophysics.)

South   North

Heated groundwater

Plug of
new magma

A

Slide block

B

North
directed
blast

C

Vertical eruption
column

D

fluidize the avalanche, and it accelerated rapidly to velocities of 250 kilometers per hour. One lobe of the gigantic mass plowed through the west arm of Spirit Lake and into the valley to the north of the lake. The momentum of another lobe carried it over a ridge 360 meters above the floor of the valley that it had just swept across, but the bulk of the fluidized debris funneled down the valley of the Toutle River for 21 kilometers, forming a hummocky deposit 1 to 2 kilometers wide and up to 200 meters thick (Figure 43).

Superheated groundwater flashed into steam as the great avalanche of rock and ice suddenly released the pressure in the volcanic edifice. At the same time, dissolved gases began exploding from the shallow magma body that had recently pushed up into Mount St. Helens. The steam blast, magma explosion, and giant avalanche combined to form a lateral blast of dense, debris-filled steam clouds as hot as 300°C, which surged northward from the breached volcano at speeds as fast as 1000 kilometers per hour. This blast of steam and its fluidized charge of volcanic rock fragments devastated nearly 600 square kilometers of mountainous terrain northwest, north, and northeast of Mount St. Helens. The ground-hugging black clouds rolled over four major ridges and valleys, extending as far as 30 kilometers from their source.

The destruction was complete. For the first few kilometers, entire trees as large as 2 meters in diameter were uprooted and swept away with the roiling explosion cloud. Beyond this was a blowdown zone 10 to 15 kilometers wide where prime fir trees were snapped off like

42 *(Facing page)* These schematic cross sections of Mount St. Helens show the early stages of the May 18, 1980, eruption: A. The shallow plug of magma injected beneath the volcano has caused the high north face to bulge outward and has heated the groundwater. Minor explosions of heated groundwater have excavated a small summit crater. B. The magnitude 5 earthquake caused by the continuing injection of the magma plug shakes loose the oversteepened north flank of the mountain into two or more massive slide blocks. C. As the superheated groundwater and magma plug are suddenly depressurized, they explode into a huge steam blast that sweeps northward for many kilometers. D. After the avalanche and blast have exposed the main magma conduit, gases dissolved in the magma continue to boil out in continuous vertical jetting explosions. (Modified from J. G. Moore and W. C. Albee, U.S. Geological Survey Professional Paper 1250, 1981, p. 132.)

43   View to the east showing Mount St. Helens and the huge avalanche deposit in the valley of the North Fork of the Toutle River. (U.S. Geological Survey photograph by R. M. Krimmel, June 30, 1980)

matchsticks (Figure 44). At the outer limits of destruction, trees were still standing, but their needles had been scorched beyond recovery.

The blowdown zone looked as if a great shock wave or concussion front had knocked down the trees in a pattern radial to the exploding peak, but the evidence showed that this was not what had happened. Survivors near the edge of the devastated area heard only a moderately loud explosion or roaring sound two to three minutes before the black cloud with its hot hurricane winds descended upon them. The average velocity of the growing front of the steam-blast clouds was below the speed of sound. On closer inspection, the pattern of downed trees showed turbulent eddies and curving streamlines. Near the edge of the devastated area, most trees were blown down in down-valley directions, even turning toward the source of the surging clouds (Figure 45).

Gravity apparently provided energy to the dense, fluidized mass as the initial steam-blast energy waned. As the turbulent internal winds abated, ash and rock debris from the dense clouds settled to the surface in deposits that ranged in thickness from about 1 meter to a few millimeters. Angular fragments of old volcanic rocks and hot but recently solidified magma were carried in the blast clouds as far as

44 Trees blown down by the hot, debris-filled hurricane winds in the Mount St. Helens steam blast cloud fell into streamlined rows over hundreds of square kilometers. These logs are large Douglas firs and other conifers 1 to 2 meters in diameter and 40 to 60 meters long. (Photograph by Katia Krafft.)

Key

Lava dome

Pyroclastic flow deposits

Mudflow deposits and
scoured areas

Scorch zone

Down–timber; arrows indicate
direction of tree blowdown

Debris-avalanche deposits

▲ Mountain summits

45   This map of the region surrounding Mount St. Helens shows the effects of
the May 18, 1980, eruption and the subsequent lava dome. The devastated area is
more than 500 square kilometers. The size of Spirit Lake was increased by the
eruption, as seen in both the old and the new shorelines. (Adapted from the U.S.
Geological Survey.)

15 kilometers from the exploding peak. Trees were charred and blackened on their sides facing the blast for 7 kilometers on the northwest and north and as far as 18 kilometers on the northeast flank of Mount St. Helens.

The worst was over in the first five minutes, but the vertical eruption column roared on, reaching a maximum height of about 26 kilometers at 9 A.M. and then diminishing in altitude to about 14 kilometers for much of the day. The source of this almost continuously exploding and uprushing column of gas and ash was the effervescing shallow magma body being progressively cored out to greater and greater depths. The abrasive uprush continued to enlarge the horseshoe-shaped crater that initially had been formed by the avalanche and lateral blast.

High-altitude winds blew all day to the northeast, and ash began falling on central Washington towns by midmorning. At Yakima (150 kilometers away), the first ashfall was a sand-size "salt and pepper" layer composed of dark rock fragments and lighter colored feldspar crystal fragments, overlain by a thicker, silt-size layer of volcanic glass particles. Thirty kilometers north of Yakima, the maximum ashfall was about 20 millimeters. Eastward the fine ash thickened, reaching more than 70 millimeters near Ritzville, Washington, 330 kilometers away from Mount St. Helens. Here the texture was like talcum powder. Although the ash thinned to 5 millimeters near Spokane, 430 kilometers northeast of the volcano, by 3 P.M. visibility was still reduced to 3 meters in near darkness. A trace of ash fell on Denver about noon on May 19; the ash cloud crossed the United States in three days.

Mudflows were another major result of the eruption. These flows consisted of a slurry of volcanic ash and fine rock particles mixed with water, which had the consistency of wet cement. The ash blanket near the mountain and the crushed rock in the avalanche deposit provided the solid matter; the extra water probably came from several sources— melting snow and ice, the water displaced from Spirit Lake and the North Fork of the Toutle River, water from the compaction and settling of the saturated avalanche deposits, and condensing steam.

The highest flood crest in the North Fork of the Toutle River arrived at the gauging station that evening and destroyed it. High mudflow marks on trees indicate a flood stage of 16 meters, 9 meters higher than any flood previously recorded on the Toutle River. Downstream, mudflow deposits clogged the channels of the Cowlitz River and caused severe shoaling of the navigation channel in the Columbia River.

Sometime after the initial avalanche and steam-blast eruption, pyroclastic flows of fine ash and pumice blocks began rushing down the north slope of Mount St. Helens, through the breach in the newly formed crater (Figure 46). These fluidized emulsions of hot (though not molten) rock fragments and volcanic gases formed dense flows that poured from the crater beneath the rising ash cloud. Successive flows churned down the north slope at speeds up to 100 kilometers per hour, covering the earlier landslide and blast deposits and reaching the south edge of Spirit Lake. Contact by these hot pyroclastic flows with water caused secondary explosions, sending steam and ash clouds 2 kilometers high.

The total disaster was staggering: fifty-seven people killed and more than $1 billion in loss and damage—mostly to the lumber industry. Perhaps the greatest damage was psychological: in the northwest United States the image of the Cascade volcanoes had changed from silent guardian peaks to potential killers (Figure 47).

Although the suddenness, complexity, and scale of the May 18 eruption were not anticipated, the fact that a major eruption took place was not a complete surprise. In 1978 Dwight Crandell and Donal Mullineaux of the U.S. Geological Survey issued a report about Mount St. Helens that concluded with the warning, "The volcano's behavior pattern suggests that the current quiet interval will not last as long as a thousand years; instead, an eruption is more likely to occur within the next hundred years, and perhaps even before the end of this century."

This forecast was based on their studies of the character, distribution, and ages of prehistoric eruption deposits from the volcano, as well as scattered accounts by explorers describing eruptions of Mount St. Helens between 1831 and 1857. Just two years after this forecast was published, the quiet interval ended.

Beginning on March 20, 1980, a swarm of small to moderate earthquakes centered beneath Mount St. Helens announced the volcano's awakening. A seismograph had already been placed on the west flank by University of Washington seismologists, and soon after the swarm started, additional seismographs were installed to record and locate the earthquakes more precisely. On March 25, within a twelve-hour period, there were forty-seven earthquakes of magnitude 3 or more at shallow depths under the volcano's north flank.

46 Computer-graphic models of Mount St. Helens before and after May 18, 1980. The avalanche and eruption removed the top 400 meters of the peak, forming a breached crater 2 kilometers wide, 4 kilometers long, and 700 kilometers deep. These views are from the northeast and exaggerate the relief. (Created by the U.S. Geological Survey and Dynamic Graphics, Inc.)

The first eruptive activity—small explosions of steam—began two days later, forming a new crater about 70 meters wide on the snow- and ice-covered summit. Large east–west cracks also developed across the summit area. A second small crater formed on March 29, and blue flames—possibly burning hydrogen sulfide gas—were visible from the air at night. The next day, scientists counted ninety-three small eruptions of steam and ash.

On April 1 the seismographs recorded the first volcanic tremor, a more or less continuous ground vibration that is observed at many active volcanoes. The precise cause of volcanic tremor is not clear, but it presumably reflects the movement of magma or the rumbling formation and collapse of subsurface gas bubbles associated with the incipient boiling of groundwater or gases dissolved in magma.

The small steam and ash explosions continued—some as single explosions and others as pulsating jets lasting for hours. While the craters grew into a single oval pit 500 meters by 300 meters across and 200 meters deep, columns of gas and ash climbed to 3 kilometers above the summit. The ash consisted of fragments of old volcanic rock, and the gas emissions were largely steam mixed with carbon dioxide, sulfur dioxide, hydrogen sulfide, and hydrogen chloride. The small explosive eruptions during this period apparently took place when groundwater high in the volcanic cone was heated above its subsurface boiling point and suddenly flashed into steam, much as a geyser does, but with enough energy to tear loose rock fragments and blast out a crater.

As the high level of seismic activity continued—about fifty earthquakes of magnitude 3 or greater per day—another ominous sign became apparent. Geologists who were on the mountain taking measurements noticed that the profile of the north flank seemed to show distortion. When they compared their view with a photograph in a climbing guide,

47   (*Facing page*) Before and after views of Mount St. Helens. The photograph at the top was taken in June 1970; the one at the bottom in May 1980. The pre-1980 summit elevation was 2950 meters, the cone rising from a base with an elevation of about 1000 meters. The rim of the posteruption crater has elevations ranging from 2400 to 2550 meters, and the crater floor elevation varies from 1800 to 1900 meters. (1980 photograph by Ross Hamilton, U.S. Geological Survey) taken from the same place, it was evident that a major deformation was taking place. Careful measurements showed that a bulge was growing upward and outward, creating a spreading network of large cracks in the cover of snow and ice. Maps made by the U.S. Geological Survey from aerial photographs taken on April 12 disclosed that the bulge was 2 kilometers in diameter and had already moved up and out by as much as 100 meters.

Repeated ground surveys made in late April and early May showed that the bulge was continuing to expand northward at a rate of about 1.5 meters per day. This rapidly deforming area was directly over the center of the earthquake zone 2 kilometers below. Scientists on the site agreed that the continuing seismic swarm and surface bulging were signs that magma was being injected at a shallow depth beneath Mount St. Helens. They concluded that if the intrusion of magma continued at this high rate, a significant eruption was likely; the main questions were how soon it might occur and how violent it might be.

Small steam and ash explosions continued through the first half of May, and the rate of seismic and deformation activity continued unabated. On May 15, 16, and 17, there were no eruptions, and although the earthquakes and bulging continued, their rates showed no change. May 18 dawned as a beautiful spring day, but at 8:32 that morning, the scientists' worst fears became a reality.

Among the dead and missing that day was David Johnston, a U.S. Geological Survey scientist who was measuring the growing bulge from a ridge 10 kilometers north of the volcano's summit. The scale of the eruption was unexpectedly large, especially the great area devastated by the lateral blast, which was unprecedented in the prehistoric eruption record of Mount St. Helens. The suddenness with which the avalanche broke open the hot groundwater and magma system beneath the summit of the volcano caused much of the violence. Had the unroofing of the shallow magma injection occurred in a slower, more piecemeal way—for example, in a series of increasingly larger explosions over a period of several days—the area devastated by the eruption would have been much smaller.

On five more occasions through the summer and early fall of 1980, smaller explosive eruptions of Mount St. Helens generated ash clouds and pyroclastic flows (Color Plate 16). Twice, domes of viscous lava were pushed up over the vent area in the crater but then were blown out by explosions. However, between late 1980 and 1986 a large lava dome episodically squeezed up over the vent area, and at the time of this writing (1997), it is about 1 kilometer in diameter and 300 meters high (Figure 48). These episodes of rapid dome growth typically lasted for several days, separated by quiet intervals of a few months to several years.

These dome-growth episodes were preceded by several days of increasing seismic activity and increasing rates of ground deformation at the margins of the dome. Predictions of the onset of rapid dome growth based on this continuous seismic and deformation monitoring were excellent. Because the amount of growth in each episode was

48   From 1981 to 1986, a dome of viscous lava intermittently squeezed up from the vent at Mount St. Helens. Seventeen episodes of dome growth, generally lasting from 1 to 3 weeks, built this 1-kilometer-wide, 250-meter-high dome of dacitic lava. More viscous than basalt, this high-silica lava piled up over its vent. (U.S. Geological Survey photograph by Lyn Topinka, 1984.)

roughly proportional to the length of the quiet interval that preceded it, the amount of lava in each new episode also was predictable.

The last episode of dome growth occurred during October 1986. If the lava production continues to be proportional to the quiet interval, then the next episode should be large. However, the rate at which magma reached the surface and near surface at Mount St. Helens slowed between 1980 and 1986, a trend that was also reflected in a generally decreasing amount of seismic activity. These waning trends and the absence of newly erupted magma over the past decade suggest that the recent eruptive period is over.

Mount St. Helens is one of fifteen major volcanoes of the Cascade Range, a mountain chain that extends from northern California into western Canada. Although each of these volcanoes is potentially active, only one other—Lassen Peak in California—has erupted in this century. The sixty-three-year repose between the 1914–1917 Lassen eruptions and the 1980–1986 eruption of Mount St. Helens was long enough for most Americans to forget that the Cascade Range is still alive and dangerous. However, since May 18, 1980, many now remember.

Sources

Carson, Rob. *Mount St. Helens: The Eruption and Recovery of a Volcano*. Seattle: Susquatch Books, 1990.

Decker, Robert, and Barbara Decker. "The Eruptions of Mount St. Helens." *Scientific American*, March 1981.

Lipman, Peter, and D. L. Mullineaux, eds. *The 1980 Eruptions of Mount St. Helens*. U.S. Geological Survey Professional Paper 1250, 1981.

Mullineaux, Donal. *Pre-1980 Tephra-Fall Deposits Erupted from Mount St. Helens, Washington*. U.S. Geological Survey Professional Paper 1563, 1996.

Pringle, Patrick. "Roadside Geology of Mount St. Helens National Volcanic Monument and Vicinity." Olympia: Washington Department of Natural Resources Information Circular no. 88, 1993.

Tilling, Robert, Lyn Topinka, and Donald Swanson. *Eruptions of Mount St. Helens*. Washington, DC: U.S. Government Printing Office, 1990.

6

Ring of Fire

49   Cerro Negro Volcano, Nicaragua, 1968.

Far away, shrouded in fog we saw the silhouettes
of the volcanic islands—links in Pluto's mighty chain.

—Y. Markhinin (1971)

America is drifting away from Europe, and it is also drifting toward the Philippine Islands. In fact, all the plates around the Pacific are slowly converging along the subduction zones that encircle this vast ocean. Volcanoes and earthquakes along these converging plate margins form a "Ring of Fire" that nearly surrounds the Pacific Basin (Figure 50).

Mount St. Helens and the other volcanoes of the Cascade Range are located on the subduction zone between the small Juan de Fuca Plate and the North American Plate. Volcanoes in Mexico, Central America, the Andes, Antarctica, New Zealand, Papua New Guinea, the Philippines, Japan, Kamchatka, and Alaska complete the circuit of the Ring of Fire. Another major belt of converging plate margins runs from Indonesia through the Himalayas and the Middle East to Greece and Italy. Where the continental edges of two plates converge, as in the Himalayas, volcanoes are rare and the earthquake zones more diffuse.

Subduction zones generate about four hundred of the world's more than five hundred known active volcanoes; that is, those that have erupted at least once in recorded history. The Mediterranean volcanoes of Italy and Greece are included in this tally.

There are major differences between subduction volcanoes and the volcanoes along oceanic rifts. *Subduction volcanoes* form island arcs and high mountain chains rather than submarine ridges; they are more explosive and produce large volumes of ash as well as lava flows; and their products are more variable in composition. Even their basic shapes as individual mountains differ from those volcanoes that form at diverging plate margins. *Rift volcanoes* are located at the exact edges of the separating plates, while subduction volcanoes occur on the overriding plate about 100 to 200 kilometers landward from the deep ocean trenches that mark the edges of the converging plates.

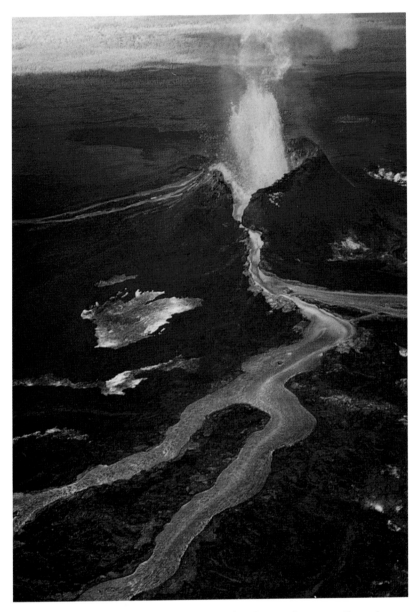

Plate 1   High lava fountains and rivers of molten rock repeatedly gush out during the episodic phase of Kilauea Volcano's Puu Oo eruption. (Photograph by J. D. Griggs, U.S. Geological Survey, January 31, 1984)

Plate 2 A 30-meter-high "curtain of fire" spurts from a long fracture at the beginning of the Puu Oo eruption. (Photograph by J. D. Griggs, U.S. Geological Survey)

Plate 3 A pahoehoe lava flow creeps across a lawn before destroying another home on its way to the sea. (Photograph by J. D. Griggs, U.S. Geological Survey)

Plate 4   An arching lava fountain, 10 to 15 meters high, is hurled from an early vent of the Puu Oo eruption. (Photograph by J. D. Griggs, U.S. Geological Survey)

Plate 5   As lava pours into the sea, the glow of the stream of red hot rock illuminates the steam cloud. (**Photograph by** Jane Takahashi, U.S. Geological Survey)

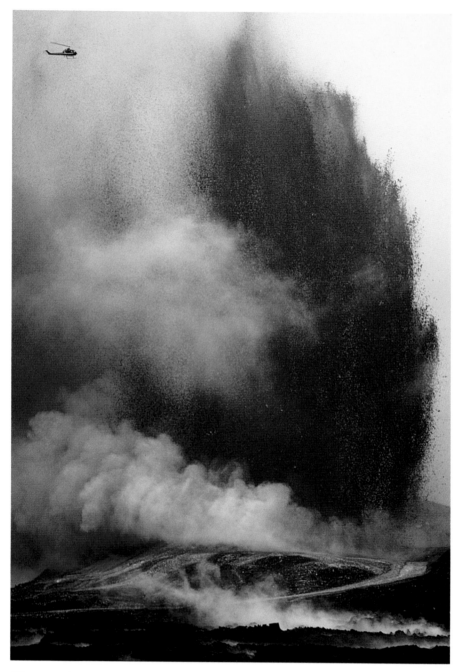

Plate 6 A 400-meter-high lava fountain during episode 31 of the eruption dwarfs a civil defense helicopter. (Photograph by J. D. Griggs, U.S. Geological Survey)

Plate 7 A fume cloud from a vigorous eruptive episode, as seen from the Hawaiian Volcano Observatory, 20 kilometers away. (Photograph by J. D. Griggs, U.S. Geological Survey)

Plate 8 The thin strands of volcanic glass spun out in a lava fountain are called *Pele's hair*. Individual strands often reach 20 centimeters in length. (Photograph by D. W. Peterson, U.S. Geological Survey)

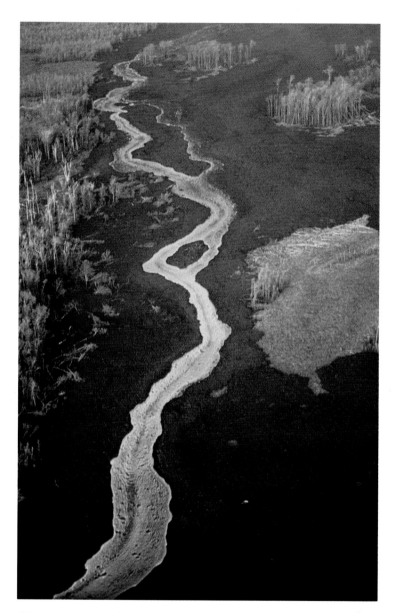

Plate 9 A river of molten lava 10 to 20 meters wide flows from Puu Oo and forms the central channel of a black aa lava flow. The trees on the left side of the flow are about 15 meters high. (Photograph by J. D. Griggs, U.S. Geological Survey)

Plate 10 The advancing front of a 2- to 3-meter-thick aa flow moves slowly but stops for no one. (Photograph by J. D. Griggs, U.S. Geological Survey)

Plate 11 Molten rock wells up in a 100-meter-wide lava lake and spills down a deep channel (left center) to feed crusted-over lava tubes. Pahoehoe flows from this continuous phase of the eruption move slowly but inexorably down the southeast flank of Kilauea toward the ocean in the background. (Photograph by J. D. Griggs, U.S. Geological Survey)

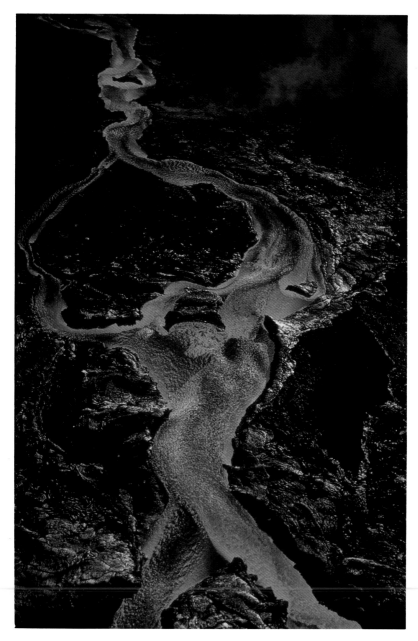

Plate 12   While photographing lava streams flowing down Mauna Loa
Volcano during its 1984 eruption, Katia Krafft captured this dramatic
moment, which she named "Pele Dancing," to honor the Hawaiian
goddess of volcanoes.

50 Ring of Fire around the Pacific Ocean, delineated by locations of earthquakes occurring between 1963 and 1977 with a magnitude greater than 4.5, as determined by the U.S. Geological Survey's National Earthquake Information Service. (Map plotted by the Environmental Data and Information Service of the National Oceanic and Atmospheric Administration.)

Chains of subduction volcanoes form graceful arcs, thousands of kilometers in length, across the globe. A close look often shows five or ten volcanoes in a fairly straight line a few hundred kilometers in length, with a slight bend, or *offset*, between adjoining groups. The bends between the groups form the overall arc, which is generally convex toward the open ocean side.

Japan is a good example of a subduction chain, with about fifty volcanoes along parts of four arcs. The deep-sea trenches lie 200 kilometers toward the Pacific Ocean side of the volcanoes and mark the actual boundary between the plates. A zone of earthquakes dips from near the Earth's surface at the trenches to below the island arcs. This region roughly outlines the edges of the Pacific and Philippine Plates as they plunge beneath Asia.

The belt of volcanoes grows on the overlying plate where the earthquake zone is 100 to 200 kilometers beneath the Earth's surface (Figure 51). Most of the large volcanoes are on the eastern edge of this belt, where the earthquake zone is shallower. Japanese geologists refer to this eastern edge as the *volcanic front*. The number of volcanoes and

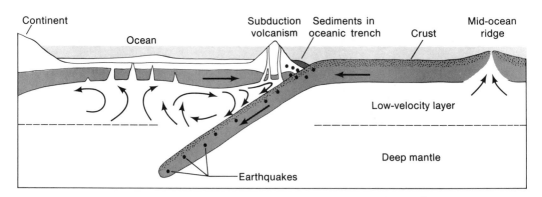

51   The plate of crust and solid upper mantle formed at the mid-ocean ridge moves on the plastic low-velocity layer and is thrust beneath a converging plate at a subduction zone. An oceanic trench forms where the plates converge, and an island arc of compressed rocks and volcanoes is built over the downgoing plate. (From M. Nafi Toksoz, "The Subduction of the Lithosphere." Copyright © 1975 by Scientific American, Inc. All rights reserved.)

their production of volcanic rocks decreases gradually with distance away from the volcanic front toward Asia, but stops abruptly on the Pacific side of the front.

This setting is generally characteristic of subduction zone volcanoes and indicates that magma is generated when the descending plate sinks to a depth of 100 to 200 kilometers. At one time geologists believed that the friction of the converging plates generated heat and that this extra heat helped melt the magma for subduction volcanoes. Numerical models of the temperatures beneath subduction zones show, however, that the underthrusting plate carries its lower surface temperature down with it, in effect cooling part of the region deep beneath the volcanic front. Most geologists now believe that the magma generated in subduction zones is formed by the fluxing action of water and calcium carbonate in the oceanic sediments dragged down on the top of the underthrusting plate. Water, carbon dioxide, and minerals that contain sodium and potassium lower the melting temperature of rock. The existing temperatures in the low-velocity layer at a depth of 100 to 200 kilometers along the top of the subducting plate are high enough to melt this combination of rock and volatiles. The addition of water and calcium carbonate also increases the amount of gas that boils out of magma as it rises toward the surface, and it explains the greater explosiveness of subduction volcanoes.

When enough magma has formed at depth, its lesser density exerts an upward force, and its ascent toward the surface may be aided by fractures in the overlying plate. The distance between volcanoes in island arcs is usually about 50 to 70 kilometers, a spacing that is about the same as the thickness of the overlying plate. In experimental models, the distance between fractures created in a rigid plate resting on a soft layer tends to be about the same as the thickness of the plate.

Figure 52 shows a good conceptual model of the island of Sumatra in Indonesia. The Australian Plate on the left is plunging beneath the continental plate of Southeast Asia on the right. Scrapings from the seafloor of the Indian Ocean pile up into the submarine accretionary wedge and the islands of deformed sediments called the *outer-arc high*. The back-arc region to the right of the active volcanoes in Sumatra is filled with oil-rich sedimentary rocks.

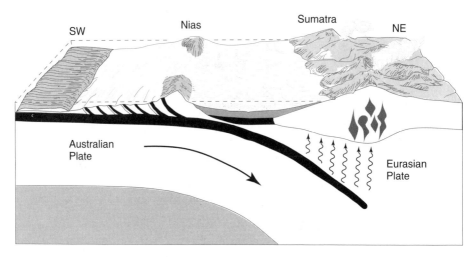

52   The convergent plate boundary in Indonesia where the Australia Plate is subducted beneath the Eurasian Plate. The outer arc forms a chain of small islands like Nias, made up of sedimentary rocks. The subduction volcanoes of Sumatra, on the inner arc, occur above where the top of the subducted plate reaches a depth of 100 to 150 kilometers. Water and carbon dioxide derived from the sediments carried down on the Australia Plate reduce the melting temperature in the upper mantle beneath Sumatra and contribute to the formation of magma. Bodies of magma then rise because they are lighter than the surrounding, unmelted rock. Some of the magma solidifies within the crust, and some escapes to the surface to form the active volcanoes. (After B. Clark Burchfiel, "The Continental Crust," *Scientific American*, September 1983.)

In some places, the subducted plate dips at a low angle; in others, at a nearly vertical angle. Christopher Scholz, a geophysicist at Columbia University, and Jaime Campos, a seismologist at the University of Chile, proposed a novel explanation for these variations. In their "sea anchor" model, the dip of the subducted slab—and therefore the nature of the volcanoes and structures on the overlying slab—depends on the relative movement of the two plates and the movement of the underlying mantle. Where the plates are pushing together, as in the Andes, the dip is shallow. Where both the plates are moving in the same direction, as in the Izu–Bonin Islands south of Japan, the dip is about 45°. And where the subducting plate is moving in the same direction as the overlying plate at nearly the same speed, as in the Mariana Islands in the Western

Pacific, the dip is nearly vertical (Figure 53). They compare these situations to the angle of a sea anchor cast from a boat where the direction of the ocean current versus the wind direction is similar to these three examples. This is an attractive, albeit simple, model, and in a complex world simplicity is most welcome.

The silica content of most rift volcanoes is about 50 percent, varying by only a few percentage points, but the silica content of subduction volcanoes varies from about 50 to 75 percent. The silica content largely determines which minerals crystallize from the cooling melt and thereby controls the type of volcanic rocks that are formed and the sediments that are derived from their erosion. For example, the mineral quartz (SiO_2) does not form during crystallization until the silica content of magma exceeds about 55 percent. In Iceland there is no quartz in the basaltic lavas, and there are no white, sandy beaches composed of quartz grains. On the Ring of Fire, however, the more silica-rich volcanic rocks and the granitic rocks beneath them often contain quartz grains that erode to form white, quartz-sand beaches.

The composition of volcanic rocks is principally controlled by three factors: partial melting, partial crystallization, and contamination by surrounding rocks. *Partial melting* is the process by which part of the solid upper mantle or lower crust of the Earth is turned into magma. Rocks are a mixture of several chemical compounds, and not all of them have the same melting points. In partial melting, the first fraction to melt will have a larger proportion of the low melting-point compounds than the parent rock. An analogy is the extraction of alcohol from hard cider by freezing and thawing. If a completely frozen block of hard cider is slowly melted, the first fraction to melt will contain a higher proportion of alcohol than does the remaining solid block.

Partial crystallization is just the opposite. As a molten rock slowly cools, the minerals with higher melting points crystallize first and the residual melt keeps changing in composition. This is analogous to slowly freezing the hard cider and thereby concentrating the alcohol in the residual unfrozen liquid.

Silica is concentrated in the low melting–point fractions, especially when water is present. Therefore, both partial melting with small quantities of water present and partial crystallization during the ascent of magma to the surface produce a liquid with a higher silica

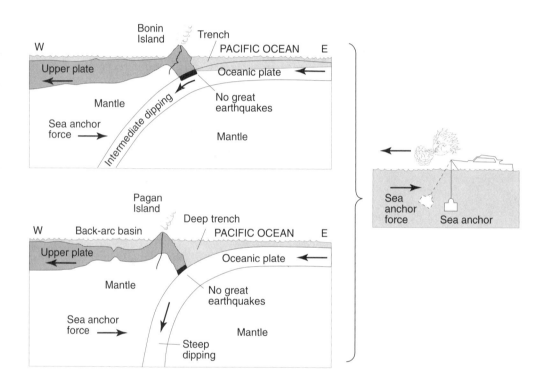

content than that of the starting material. In general, then, magmas are richer in silica than are their upper mantle sources and may be further "distilled" during crystallization to produce even more silica-rich rock.

Magma rises greater distances in subduction volcanoes than it does in rift volcanoes. Also, the magma that supplies subduction volcanoes

53 (*Facing page*) Christopher Scholz at Columbia University and Jaime Campos at the University of Chile developed this cartoon to explain variations in the dip and character of subduction zones. Their sea anchor concept compares the relative motion of a boat to the sea, and the resulting angle of the rope to the sea anchor, to the relative motion of the upper plate in a subduction zone (the wind) to the Earth's mantle (the sea). In the top example, the Nazca Plate (oceanic) moves east, and the South America Plate (Chile) moves west. The sea anchor force pushes east with the descending slab, resulting in a shallow-dipping subduction zone. This in turn causes a wide area of high friction between the two plates, large earthquakes, and a wide separation between the trench and the belt of volcanoes. In the middle example, south of Japan, both the Philippine Plate (upper) and the Pacific Plate (oceanic) are moving west. The sea anchor force pushes against the subducting plate, causing it to dip more steeply. The resulting friction between the plates is less; the number of large earthquakes is smaller; and the separation between the trench and the volcanic belt is smaller. In the bottom example, across the Northern Mariana Islands, the rearward part of the Philippine Plate (upper) is moving faster. The increased sea anchor force pushes the subducting slab to a nearly vertical dip, and the distance between the trench and the volcanic belt is small. A back-arc basin forms west of the subduction volcanoes to accommodate the difference in motion between the arc/trench and the rearward part of the Philippine Plate. (After C. H. Scholz and J. Campos, "On the Mechanism of Seismic Decoupling and Back Arc Spreading at Subduction Zones," *Journal of Geophysical Research* 100 [1995]: 103–115; and *EOS*, December 5, 1995.)

must often ascend through ancient continental rocks, already high in silica, which greatly increases their chance of *contamination*.

These concepts of partial melting, partial crystallization, and contamination provide general explanations of the origin of the many diverse volcanic rocks. Several more specific explanations have also been proposed, and the arguments favoring one process over another wax and wane on subtle details. The problem is not a lack of explanations but too many explanations. In fact, more than one process of variation is probably at work. Truth is often a mixture of conflicting opinions.

The greater proportion of water and carbon dioxide dissolved in the magma of subduction volcanoes, coupled with slightly lower eruption temperatures than in rift volcanoes, leads to more explosive eruptions. The gas wants out, and the cooler and therefore more viscous magma is holding it back, resulting in explosive eruptions of volcanic ash. Subduction eruptions often begin with explosive showers of ash and

54   Mount Fuji, or Fuji-san, the archetypal volcano. The classic shape of this nearly perfect cone with its graceful concave slopes—and its location near Tokyo—make it the world's best-known volcano. It rises to 3776 meters from a 30-kilometer-diameter base almost at sea level. Fuji-san embodies the beauty, majesty, and power of all nature. (Woodblock print by Hokusai [1760–1849] from his famous series *Thirty-Six Views of Mount Fuji*.)

lumps of pumice and end with thick, viscous lava flows. Such alternations of volcanic ash and lava flows build beautiful steep cones like Mount Fuji, the archetype of classic volcanic form (Figure 54).

About forty subduction volcanoes erupt each year. A few, like Stromboli in the Mediterranean, have been erupting nearly continuously for centuries. Others, like Pinatubo Volcano, have erupted only once in recorded history. In general, the longer the period between eruptions, the greater the likelihood that the next eruption will be a big one.

The products of subduction volcanoes are largely explosive: ash, pumice, cinders, blocks, and molten lava bombs. This is in sharp contrast to the rift volcanoes along the ocean ridges, whose products are mainly effusive (nonexplosive) lava flows. The more notorious of the world's volcanoes belong to the subduction clan: Krakatau and Mount St. Helens, as described in the previous two chapters, as well as Vesuvius, Mont Pelée, Katmai, Bezymianny, El Chichón, Nevado del Ruiz, and Pinatubo.

Vesuvius buried Pompeii in A.D. 79 and provided a time capsule of Roman art, architecture, and artifacts for its nineteenth- and twentieth-century excavators (Figure 55). Before its violent eruption in Roman times, Vesuvius was considered extinct, but it has erupted many times since then, the latest in 1944.

Mont Pelée in the West Indies annihilated the 28,000 inhabitants of the port city of St. Pierre on May 8, 1902. A pyroclastic flow—what

55 Pompeii was buried beneath 6 meters of volcanic ash for nearly 2000 years. Its destroyer and preserver, Mount Vesuvius, is in the background. Airfall ash and pyroclastic flows buried the town in one day in A.D. 79. Pliny, a Roman historian, described the eruption cloud as looking like a pine tree; a puzzling comparison until you note the tree behind the ruins.

French geologists call a *nuée ardente* (glowing cloud)—rushed down the mountainside, and the hot gases and ash bowled over and burned everything in their path, including ships in the harbor. The disaster lasted only a few minutes and left just two survivors in Saint Pierre.

Katmai, on the Alaskan Peninsula, erupted for two days in June 1912. The ashfall totaled 20 cubic kilometers and nearly buried Kodiak Island. The summit of Mount Katmai sank into a caldera 5 kilometers wide, and 10 cubic kilometers of pyroclastic flow deposits filled a valley 20 kilometers long to form the Valley of Ten Thousand Smokes. To date, the eruption of Katmai was the largest in this century.

Bezymianny, on the Kamchatka peninsula of Siberia, erupted for the first time in recorded history on October 22, 1955, culminating in a giant explosive eruption on March 30, 1956. The main eruption cloud reached 35 kilometers in height; large rocks were hurled 25 kilometers away; a caldera 2 kilometers wide was formed; and pyroclastic flows dumped 3 cubic kilometers of searing hot deposits that Russian geologists call the Valley of Ten Thousand Smokes of Kamchatka (Figure 56). Fortunately, the areas surrounding both Katmai and Bezymianny were almost uninhabited, and no one was killed. Such eruptions in Japan or Central America—or anywhere on the Ring of Fire that is densely populated—would be catastrophic.

A case in point is the giant explosive eruption of Pinatubo Volcano in the Philippines in 1991. More than 100,000 people lived in the surrounding area, but fortunately most of them were able to evacuate before the climactic eruption, which is described in Chapter 18. But a volcanic eruption does not have to be large to cause a catastrophe. Mudflows from the relatively small eruption of Nevado del Ruiz Volcano in Colombia in 1985 killed more than 20,000 people, a tragic story that is told in Chapter 10.

In prehistoric times, extremely large eruptions deposited enormous pyroclastic flows of pumice and ash covering thousands of square kilometers at several places around the Ring of Fire. Volumes of magma more than 100 times greater than that hurled from Krakatau were spewed out in single eruptions. The message of such gigantic eruptions is not imminent doom; they do not occur that often. To us their message is attention and respect; since the millions of people who reside on the Ring of Fire must live with these volcanic eruptions, it makes sense to try to understand them better.

56 The giant explosion cloud from Bezymianny Volcano, Kamchatka, March 30, 1956. The lower edge of the huge, fan-shaped ash cloud is about 7 kilometers high, and the top is about 35 kilometers high. (Photograph by I. V. Erov from 45 kilometers west of the volcano.)

Sources

Bebout, G. E., D. W. Scholl, S. H. Kirby, and J. P. Platt, eds. *Subduction: Top to Bottom*. American Geophysical Union Geophysical Monograph no. 96, 1996.

Harris, Stephen. *Fire Mountains of the West*. Missoula, MT: Mountain Press, 1987.

Hawkesworth, C. J., K. Gallagher, J. M. Hergt, and F. McDermott. *Mantle and Slab Contribution in Arc Magmas*. Palo Alto, CA: Annual Reviews of Earth and Planetary Science, 1993.

Simkin, Tom, and Lee Siebert. *Volcanoes of the World*. 2nd ed. Washington, DC: Smithsonian Institution Press, 1994.

Van Rose, Susanna, and Ian Mercer. *Volcanoes*. 2nd ed. London: British Museum, 1991.

7

Kilauea, Hawaii

57 Kilauea Iki, 1959. (Photograph by J. P. Eaton, U.S. Geological Survey.)

The marriage of Pele, goddess of earth and fire, and Kamapuaa, god of water
was short and violent. In a rage she routed him from her crater of fire
and chased him with streams of lava into the sea.

—Hawaiian Legend

K ilauea Volcano on the island of Hawaii, known also as the Big Island, is the most thoroughly studied volcano in the world (Figures 58 and 59). It is also one of the most active, with many eruptions and glowing lava lakes that stir and boil for years at a time (Color Plate 11). Almost all of Kilauea's eruptions are relatively quiet outpourings of fluid lavas. The word *quiet* can be misleading, though, for the vents often spray roaring fire fountains of incandescent lava several hundred meters into the air (Figure 57), which fall back into lava ponds or feed lava flows. These outpourings of lava are "quiet" only in contrast to explosive volcanic eruptions, which produce fragmental debris and look more like huge detonations of dynamite.

Effusive eruptions of the Hawaiian type are comparatively safe to study at close range, and scientists at the Hawaiian Volcano Observatory have been at it since 1912. We were fortunate to be studying volcanoes in Hawaii during the early 1980s when major eruptions of both Kilauea and Mauna Loa Volcanoes took place.

The new year of 1983 saw the start of the longest and largest rift eruption of Kilauea since written records began in 1823. By 1996 the eruption, which showed no sign of stopping soon, had poured forth 1500 million cubic meters of lava (1.5 cubic kilometers)—enough to build a pyramid 1.6 kilometers (1 mile) high. These flows had traveled 11 kilometers to the sea, covering 100 square kilometers of land and adding 2.2 square kilometers of new land to the island. One person was killed when a lava bench collapsed into the ocean; 181 homes and a National Park Visitor Center were destroyed; 8 miles of highway were covered by lava flows; and damage estimates exceed $60 million.

January 1, 1983, was calm, but that was soon to change. Not long after midnight, the tremor alarm triggered phone calls to staff members

58   The island of Hawaii, also known as the Big Island, is built of five volcanoes. The elevations in map A are in meters. The rift zones shown in gray on map B are zones of weakness from which numerous flank eruptions have originated. Kilauea is erupting at the time of this writing (1997): Mauna Loa last erupted in 1984; and Hualalai, in 1801. Mauna Kea's latest eruptions were about 4000 years ago, and Kohala's about 60,000 years ago. Loihi is a submarine volcano. It is presumed to be active because it has recurring earthquake swarms and because hot springs and young lava have been observed on Loihi from submersibles. (After D. W. Peterson and R. B. Moore, U.S. Geological Survey Professional Paper 1350, 1987, p. 151.)

59   Most of the lava erupted from Kilauea since 1955 has come from the East Rift Zone. Two eruptions (1971 and 1974) took place on the Southwest Rift Zone, and twelve in the summit caldera area. The flows labeled 1983–1996 are from the Puu Oo eruption. (Robin Holcomb and Christina Heliker, U.S. Geological Survey.)

of the volcano observatory. Hurrying to the observatory, which is perched on the rim of Kilauea's 3-by-5-kilometer-wide caldera, they soon found all three signs (volcanic tremor, earthquake swarms, and summit deflation) of an impending eruption.

Volcanic tremor, a continuous low-frequency vibration of the ground, was being recorded on seismographs near the East Rift Zone of Kilauea. Tremor occurs when magma is moving rapidly through underground conduits. Seismographs also showed *earthquake swarms* at shallow depths beneath the east rift, indicating that new fractures were forming, and tiltmeters registered *summit deflation*. (A tiltmeter works like an extremely sensitive carpenter's level and can detect changes in slope of 1 part per million—equivalent to a 1-millimeter rise of the end of an imaginary level that is 1 kilometer long.)

Tremor, earthquake swarms, and summit deflation all indicated that magma from a storage zone beneath Kilauea Caldera was being injected into the east rift (Figures 60 and 61). These shallow intrusions into the weak rift zones on the flank of the volcano are not uncommon; they may happen several times a year. Some reach the surface and erupt; others do not. The same question was on everyone's mind: would this be another aborted alarm, or would this shallow intrusion become an eruption?

All day on January 2 the tremor, earthquakes, and subsidence continued. The earthquakes were clustered on the middle east rift, about 10 to 20 kilometers east of Kilauea's summit. Finally, after midnight on January 3, fountains broke out in Napau Crater, which lies 15 kilometers from the summit on the East Rift Zone. By morning, fissures in a broken line 6 kilometers long were erupting in a "curtain of fire" in a remote, inaccessible forest. During the next twelve days, various vents along this fissure system erupted intermittently, but only geologists and TV crews in helicopters were able to see the spectacular start of this long-lasting eruption.

There had been a lengthy prelude to this outbreak of lava, and many clues had come from careful observations of small earthquakes and deformation of the ground surface. Kilauea's magma rises from a depth of 60 kilometers or more and accumulates temporarily in a reservoir a few kilometers beneath the summit. As the underground reservoir inflates, the pressure of the magma increases until the rocks above or on the sides of the reservoir crack and fail. Molten rock then pushes into these fractures, erupting at the surface (Color Plate 2) or halting underground as a shallow intrusion into the rift zones on the volcano's flanks. There are two persistent zones of weakness in the sides of Kilauea: the East Rift Zone and the Southwest Rift Zone.

As an injection of molten rock moves out into one of the rift zones, earthquake swarms outline the moving magma's progress. As the reservoir deflates, the summit subsides. Magma moving rapidly underground also causes volcanic tremor. By studying seismographs from many locations

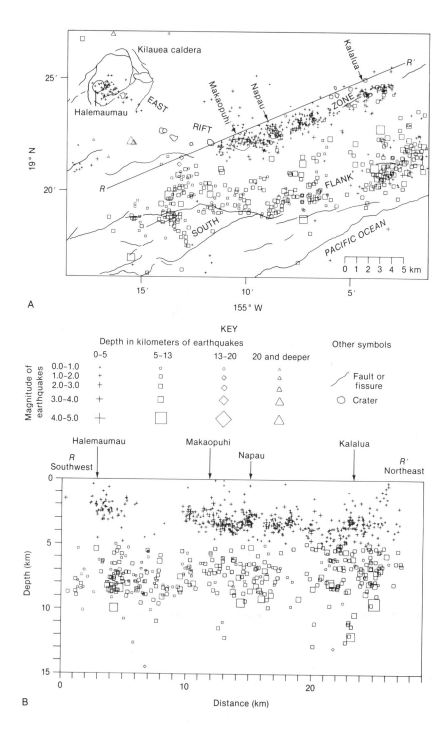

KEY

Depth in kilometers of earthquakes

Other symbols

Magnitude of earthquakes

	0–5	5–13	13–20	20 and deeper
0.0–1.0	+	·	◦	△
1.0–2.0	+	□	◇	△
2.0–3.0	+	□	◇	△
3.0–4.0	+	□	◇	△
4.0–5.0	+	□	◇	△

⟋ Fault or fissure

◯ Crater

60 *(Facing page)* A. Locations of earthquakes beneath the summit and the East Rift Zone of Kilauea during January 1983. The shallow quakes between Makaopuhi and Kalalua were caused by the magma forcing its way into fractures at the start of the Puu Oo eruption. B. Earthquake locations projected into the cross section show that the magma conduit had its axis at a depth of about 3 kilometers. (From E. W. Wolfe et al., U.S. Geological Survey Professional Paper 1350, 1987, pp. 482–483.)

KEY

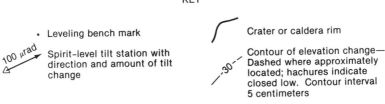

· Leveling bench mark

100 μrad Spirit-level tilt station with direction and amount of tilt change

Crater or caldera rim

-30- Contour of elevation change— Dashed where approximately located; hachures indicate closed low. Contour interval 5 centimeters

61 The summit of Kilauea subsided as magma drained into the east rift to supply the start of the Puu Oo eruption. The level surveys that determined this change were made between September 27 and October 5, 1982, and repeated between February 7 and 22, 1983. A continuously recording tiltmeter on the northwest side of Kilauea Caldera showed that nearly all this subsidence occurred between January 2 and 7, 1983. (From E. W. Wolfe, et al., U.S. Geological Survey Professional Paper 1350, 1987, p. 488.)

on the volcano and measuring small changes in uplift, subsidence, expansion, or contraction of the volcano's surface, these underground movements of magma at Kilauea are routinely detected and mapped.

There were twelve intrusions of magma beneath the east rift from 1978 through 1982, only two of which erupted to the surface, one in 1979 and a very small event in 1980. Besides these rapid intrusions into new underground cracks, slower leakage from the summit reservoir into existing fractures caused additional swelling of the east rift in the area where the 1983 eruption fissures opened. This region was ready for a major eruption.

The fissures erupted intermittently between January 3 and 23; then lava emission stopped and the summit of Kilauea began to reinflate. Almost 5 square kilometers of land had been covered with 14 million cubic meters of basaltic lava flows. Because short-lived volcanic events are common in Hawaii, many people thought the eruption was over, but on February 10, minor lava emission resumed, and from February 25 to March 4, vigorous lava fountains spewed from a vent about two-thirds of the way down the original fissure system. Again the eruption stopped, and tiltmeters at the summit, which had indicated subsidence, reversed direction and showed reinflation.

When summit inflation recovered to a critical level, would the eruption resume again? The answer was yes. During the third episode, lava gushed forth from March 28 to April 9, and in the fourth episode, at a new site uprift near the center of the original fissure system, lava fountains erupted from June 13 to 17. Between July 1983 and July 1986, the general eruption pattern—short periods of high lava fountains from the same vent area lasting a few hours to days, separated by longer repose periods averaging about one month—was repeated an astounding forty-three more times (Color Plates 1 and 6).

Although this three-year period of activity began to seem repetitious, geologists in charge of field observations discovered interesting trends and changes in eruptive habits that helped reveal the inner workings of Kilauea Volcano. Figure 62 shows both the cumulative volume of lava produced by the first twenty eruptive episodes, and the subsidence–inflation cycles of the summit magma chamber. Notice how much steeper the inward tilt of the summit is during an eruption episode than is the outward tilt during reinflation. This indicates that

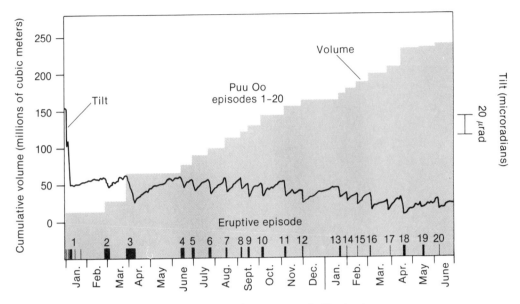

62   From 1983 until 1986, Puu Oo erupted intermittently. During its repose periods, the summit tilt showed a slow uplift, and during episodes of vigorous lava fountaining (shown in black with the episode number), the tilt indicated rapid subsidence of Kilauea's summit area. The substantial tilt loss at the beginning of the eruption (episode 1) was caused by the large volume of magma needed to fill the fracture system that initiated the eruption. (From E. W. Wolfe, et al., U.S. Geological Survey Professional Paper 1350, 1987, pp. 490–491.)

the rate at which lava was erupted was much faster than the rate of supply of magma from depth.

The average rate of lava emission during the episodic phase of eruption was 160 million cubic meters per year (Figure 63). As the magma erupted, most of the volcanic gases boiled out in vigorous fountains, although some gas stayed trapped in bubble holes, called *vesicles*, in the solidifying lava. To calculate the magma volume, subtract the volume of the vesicles—about 25 percent. The average rate of magma erupted in the early part of the eruption was about 120 million cubic meters per year. Later in the eruption, the rate of lava emission slowed and then returned to its earlier levels. To date, the average rate of magma production has been close to 100 million cubic meters per year. Although

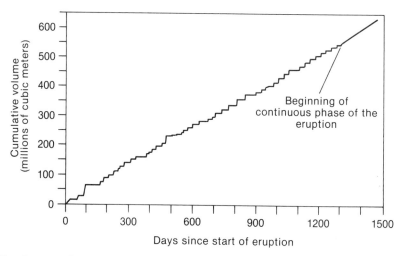

63   Lava production rate of the Puu Oo eruption, January 1983 to December 1986. Lava production rate during the Puu Oo eruption from 1983 to 1989 was remarkably steady. In the episodic phase of the eruption, each fountaining event produced about 13 million cubic meters of lava. Each of the repose periods between episodes lasted about a month. After the continuous phase began in January 1986, the slow but steady emission of lava amounted to about 0.4 to 0.5 million cubic meters per day. (From G. E. Ulrich, U.S. Geological Survey.)

the net tilt change during the past fourteen years shows that the summit has subsided, the amount of magma withdrawn from beneath the summit has been small compared with the total amount of magma erupted. Nonetheless, the supply of new magma from depth has largely kept up with the amount of lava erupted. This balance between the eruption rate and the resupply of magma implies that the eruption could be very long lasting.

As the early phase of the eruption progressed, the fountaining episodes became shorter, but the rate of lava emission increased (Figure 64), maintaining roughly the same amount of lava produced per episode (about 13 million cubic meters). As the discharge rates increased, the lava fountains grew higher, exceeding 400 meters during episodes 24 to 30. This increase in vigor may have been partly caused by improved plumbing. That is, the magma coursing underground during each episode of eruption probably enlarged the conduits. Magma is hot enough (about 1150°C) to melt the wall rocks and enlarge the channels through which it moves. This is also thought to be the reason that a fissure eruption usually becomes localized at a single vent as a Hawaiian eruption progresses: where a fissure is narrow, less lava can flow through it, and the crack tends to heal shut as some lava cools and clings to the wall rock. Wider places in the fissure are enlarged as hot magma rapidly surges through them; hence, they increase in width and capture more and more of the flow.

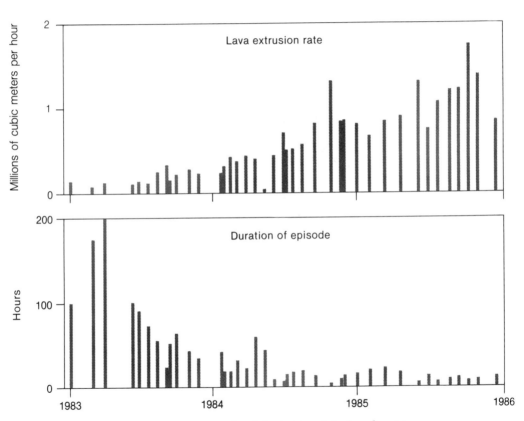

64   The rate at which lava was erupted and the height of the lava fountains generally increased as the Puu Oo eruption progressed, although the duration of each episode generally shortened. The net effect was to maintain a roughly equal amount of lava (about 13 million cubic meters) erupted in each episode. (From G. E. Ulrich, U.S. Geological Survey.)

Lava temperatures near the vents increased as the eruption progressed, from about 1120° (episode 1) to 1140°C (episode 18), and the magnesium content of the basalt increased from about 6 percent to 7.5 percent. These changes are interpreted to mean that some of the magma erupting in the early episodes had been stored in the rift zone before eruption, cooling and precipitating olivine, a mineral rich in magnesium. Later in the eruption, most of the lava emitted was from hotter, more primitive magma stored and replenished in the summit reservoir.

With the vent localized in one spot along the fissure (episodes 4 through 47), the eruption began to build a major cinder and spatter cone around it (Figure 65). The cone was 20 meters high after episode 4,

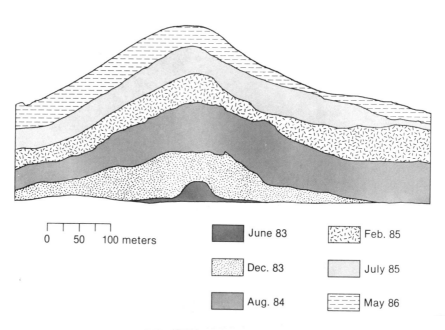

0   50   100 meters

June 83   Feb. 85

Dec. 83   July 85

Aug. 84   May 86

65   The growth of Puu Oo's cinder and spatter cone. Episodes 4 to 47 (1983–1986) were erupted from this vent and built the cone to a height of 255 meters with a base about 1 kilometer wide. These profiles are looking up the rift zone and are not vertically exaggerated. Since 1986, the growth has stopped, and slumping has greatly modified the cone's shape. (From Christina Heliker, U.S. Geological Survey.)

130 meters high following episode 20, and by 1986 had reached a height of 255 meters, with a base about 1 kilometer in diameter. This imposing cone became a landmark visible from roads 30 kilometers away and was given the Hawaiian name Puu Oo, "Hill of the Oo Bird." Since 1986, the cone has been slumping into a lava lake inside its crater, thus losing elevation.

The lava falling back from the high fountains formed aa flows that moved away from the vent to the northeast, east, or southeast for 5 to 8 kilometers. Flows of this type are generally about one-quarter to one-half kilometer wide and 3 to 5 meters thick, fed by a central river of incandescent lava about 10 to 20 meters across. The central stream moves quite rapidly—10 to 20 kilometers per hour—but as it spreads beneath the dark rubbly blocks near the front of the flow, it slows dramatically. The overall advance rates of the Puu Oo aa flows (aa flows have rough surfaces of broken lava blocks) were only a few hundred meters per hour. Although people had plenty of time to evacuate, several homes were crushed and burned as the relentless aa flows covered parts of a rural subdivision 7 kilometers southeast of the vent.

The reason that the early eruptions at Puu Oo were episodic and geyserlike is not clear. Two factors seem to have been involved. First was some sort of off-on valve in the conduit between the summit magma reservoir and the vent, which was sensitive to pressure changes. As the summit inflated and the magma pressure increased, new magma pushed through the conduit; then, as the summit deflated and the pressure dropped, the magma stopped flowing.

The second factor was the gas content of the magma. Lava appeared in the vent for several days before the fountaining started, and slow overflow of the vent preceded the high fountaining. Apparently a dense, gas-poor plug of lava had to overflow before the lighter, gas-rich lava that propelled the fountains could begin boiling. When the fountaining removed the top of the magma column, it reduced the pressure on the magma below, and it too flashed into boiling effervescence. This discharge continued at a high rate until the available magma and gas were exhausted.

The gases dissolved in magma are mainly water, carbon dioxide, and sulfur dioxide. At Kilauea they amount to less than 1 percent, by weight, of the magma. However, at high temperatures and low surface pressures, these gases vigorously boil out of the magma, and the expanded volume of gas is many times the volume of the molten lava, driving the high lava fountains.

The high lava fountains produced large amounts of pumice and Pele's hair. *Pumice* is a quickly cooled volcanic rock puffed up by the expanding volcanic gases in tiny bubble holes; in a sense, it is solid-ified lava foam. The ratio of gas bubble holes to glassy volcanic rock is so high that some pumice lumps are as light as Styrofoam. Fine strands of volcanic glass that spin off from molten drops of lava whirling in the turbulent lava fountains are called *Pele's hair* (Color Plate 8). Both pumice lumps and Pele's hair were often blown several kilometers downwind from the lava fountains. Heavier lumps of solid-ified lava, called *cinders* or *scoria*, and still-molten lumps of spatter accumulate near the bases of the fountains and build the cinder and spatter cone around the vent. The Puu Oo cone grew highest on the southwest rim of the vent because the prevailing northeast trade winds piled most of the fallout from the lava fountains on the downwind side.

Puu Oo's seventeenth eruptive episode brought an answer to a long-standing scientific question. For years volcanologists had specu-lated on what the underground connections of Kilauea and nearby Mauna Loa volcanoes might be and whether the activity of one would affect the other. Mauna Loa began a major eruption on March 25, 1984, and by March 30, a large flow of lava from a vent at 2900 meters elevation was headed toward the city of Hilo (Color Plate 18). Puu Oo was expected to erupt in late March, and everyone waited to see if Mauna Loa's eruption would affect the activity at Kilauea, or vice versa. Episode 17 arrived on schedule with 100-meter-high fountains that lasted for twenty-three hours and poured out 10 million cubic meters of lava. It was rare and marvelous to see both volcanoes erupting at the same time.

Even though the vent at Puu Oo was more than 2000 meters lower than the vent at Mauna Loa, neither eruption had any apparent effect on the other. Although both Kilauea and Mauna Loa Volcanoes may get their magma from the same general deep source, it is clear that there is no direct hydraulic connection between the shallow magma reservoirs of the two volcanoes.

Three and a half years after the start of the Puu Oo eruption, episode 48 in July 1986 brought a significant change. Fractures opened both uprift and downrift from Puu Oo; a new vent became established about 3 kilometers downrift and activity was now focused there. Smooth-surfaced lava flows called *pahoehoe* issued slowly and steadily from this new vent, building a broad, low pile of solidified lava around it and ponding the upwelling magma into a 100-meter-wide active lava lake (Color Plate 11). Lava sometimes overflowed the rim of the pond, but more often it escaped through a tunnel to feed pahoehoe flows that slowly advanced through lava tubes beneath their surfaces.

In contrast to the earlier episodic pulses, the downrift eruption became nearly continuous, pouring out lava at a slowly decreasing rate for more than six years. The vent and lava shield built up around it during this second phase of the Puu Oo eruption was named Kupaianaha, the Hawaiian word for "mysterious." Although the pahoehoe lava flows from this continuous phase of the eruption advanced more slowly than did the aa flows from the earlier intermittent eruptions, they slowly but inexorably moved down the southeast flank of the rift to the sea. The flows reached the ocean in November 1986, destroying rural homes and covering a highway in their path. Many more homes and a National Park Visitor Center were destroyed in later flows from Kupaianaha.

After a decreasing output of lava, this second phase of the eruption came to an end in early 1992. Saddened by the destruction, most people in Hawaii hoped the eruption was over for good, but that was not to be the case. About a week later, a new vent opened on the uprift flank of the Puu Oo cone. Nearly continuous flows from this third phase of the eruption again moved slowly toward the sea, inundating new areas on the west side of the still-growing flow field (Figure 66). A lava lake formed deep inside the old crater of Puu Oo, and its surface level now fluctuates with the emission rate of lava from the new vent area.

The molten rock moves in lava tubes beneath the surface of pahoehoe flows from the vent area to the sea, a distance of about 7 kilometers. Sometimes the tubes become blocked with chunks of lava crust. When this happens, new flows break out from holes in the lava tubes, called *skylights*, uphill from the blockage. Often the tubes clear themselves, and the flows to the sea are reestablished. The lava-generated steam sometimes causes small explosions as the red hot rock slips into

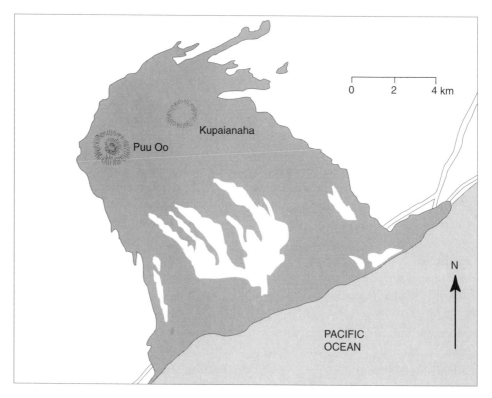

66   Lava flows from the long-lasting Puu Oo eruption on the east rift of Kilauea Volcano. From January 1983 to January 1997, 96 square kilometers of land (7 percent of Kilauea) has been covered by flows, mainly from the Puu Oo and Kupaianaha vents. The total volume of lava is nearly 1.5 cubic kilometers, and the flows have destroyed 181 homes and a National Park Visitor Center. Sixteen kilometers of road along the coast have been covered, and more than 2 square kilometers of new land has been added to the island of Hawaii. (Drawing by Richard Hazlett.)

the sea, but usually a pillar of steam is all that marks a remarkably docile meeting of fire and water (Figure 17).

As the molten lava is suddenly quenched by the sea, some of it shatters into sand-size fragments. This new sand is swept along by shoreline currents and forms black sand beaches where once the surf pounded on bare rock ledges. These new beaches accumulate in embayments along the coast, well beyond the edges of the flows entering the sea.

The billowing surface of the pahoehoe flows that form most of the land that has been added to the Big Island looks like easy walking, but its appearance is deceptive. Not only is it possible to break through the surface into a lava tube, but large sections of this newly added land can suddenly slump into the sea. This happens without warning and poses a major danger to volcano watchers trying to get a better view of the molten lava meeting the ocean. The one person who has been killed by the Puu Oo eruption was a photographer who was swept away when one of these benches violently sank beneath the waves.

Why did the Puu Oo eruption become slow and steady after the first phase? Part of the explanation may be that most of the volcanic gas was and still is venting from the Puu Oo cone and not enough gas remains in the magma at the erupting vent to drive the high lava fountains.

The gases also seem to be responsible for another phenomenon, a haze of volcanic fume locally called *vog* (volcanic smog) that has plagued parts of the Big Island since the lava began entering the ocean in late 1986. The ingredients of this vog come not only from the sulfur gases from Puu Oo and the active vent areas but also from an aerosol of hydrochloric acid droplets formed in the steam plume where the lava meets the sea. The trade winds generally blow the vog out to sea near the eruption, but it usually eddies back on the west side of the island—much to the irritation of residents used to warm, clear breezes and unobscured views of the sunset.

Although the Puu Oo eruption is the longest-lasting and largest rift eruption in recorded history, it is not unprecedented. Summit eruptions of Kilauea were nearly continuous from 1823 to 1924, and one prehistoric east rift eruption appears to have been much more voluminous—and probably longer lasting—than the present output. Each new eruption leads to important insights into how volcanoes work. But as quickly as old questions are answered, new ones arise. Why does the rate of lava emission vary? How do the lava tubes form, block up, and reopen? Why does some of the new land added to Hawaii in this eruption slump into the sea in sections larger than a football field? How does the hydrochloric acid aerosol form in the steam plumes where the lava enters the sea? There is much for the current group of scientists at the Hawaiian Volcano Observatory to learn. There always will be.

Sources

Decker, Robert, Thomas Wright, and Peter Stauffer, eds. *Volcanism in Hawaii*. 2 vols. U.S. Geological Survey Professional Paper 1350, 1987.

Macdonald, Gordon A., Agatin T. Abbott, and Frank L. Peterson. *Volcanoes in the Sea*. Honolulu: University of Hawaii Press, 1983.

Mattox, Tari N., Jeff Sutton, and Tamar Elias. "Where Lava Meets the Sea, and Volcanic Gases Create Air Pollution." *Earthquakes and Volcanoes* 24 (1993).

Tilling, R. I., and J. J. Dvorak. "Anatomy of a Basaltic Volcano." *Nature*, May 13, 1993.

Tilling, Robert, C. Heliker, and T. L. Wright. *Eruptions of Hawaiian Volcanoes: Past, Present, and Future*. U.S. Geological Survey, 1987.

8

Hot Spots

67   Lava lake in Hawaii. (Photograph by D. H. Richter, U.S. Geological Survey.)

T he close connection between the edges of the moving plates of the Earth's crust and volcanoes is emphasized in earlier chapters. Why, then, is the island of Hawaii, with Kilauea and Mauna Loa, two of the world's most active volcanoes, located right in the middle of the Pacific Plate (Figure 68)? Only 5 percent of the world's active volcanoes are located within plates, but even so, there must be some reason for their existence.

There are some obvious differences between the Hawaiian volcanoes, which form a linear belt or chain, and volcanoes located on plate margins. Topography is one. A profile section across the Hawaiian chain is completely different from a cross section of a spreading ridge or a subduction zone (Figure 69).

Another major difference is the age distribution of Hawaiian volcanoes compared with that of plate-margin volcanoes. For more than 150 years, geologists have recognized that the islands in the Hawaiian chain are older to the northwest. Since 1800, eruptions have occurred only on Hawaii, at the southeast end of the chain; the islands to the northwest are lower and more eroded. This island chain extends to Midway Island, 2500 kilometers northwest of the Big Island of Hawaii, where only a coral atoll cap covering a submerged volcanic peak marks the largely submarine ridge. The ages of the volcanic rocks of the Hawaiian Islands, obtained by radiometric dating, confirm this picture. All the rocks on the Big Island are less than 1 million years old; most of the rocks on Oahu, the location of Honolulu and Pearl Harbor, are 2 million to 3 million years old; and the ancient lava flows on Kauai, northwest of Oahu, are about 5 million years old (Figure 70).

This pattern of age distribution of Hawaiian volcanic rock is strikingly different from that of plate-margin volcanoes. In subduction zones and on mid-ocean ridges, the young volcanoes form in lines near and

68   The Hawaiian Islands rise from the southeast end of a 6000-kilometer-long
dogleg chain of seamounts. The portion north of the bend is called the Emperor
Seamounts; that south of the bend is called the Hawaiian Ridge. (Map from
World Ocean Floor Panorama by Bruce C. Heezen and Marie Tharp, initiated and
supported by the Office of Naval Research. © Marie Tharp, 1977. Reproduced by
permission of Marie Tharp. All rights reserved.)

along the margins of the plates, and on the mid-ocean ridges there is a
progressive aging of the volcanic rocks on either side of the active crest.
The Hawaiian chain is just the opposite: the volcanoes are older along
the crest of the chain instead of across the crest of the ridge.

69 Comparison of topographic profiles of subduction zones, mid-ocean ridges, and Hawaii. The vertical scale is exaggerated forty times. The island arcs and compressional mountains are basically asymmetrical, whereas the mid-ocean ridges and Hawaii are symmetrical around a central axis. The width of the mid-ocean ridges is caused by seafloor spreading.

70 The ages of islands in the Hawaiian chain in millions of years. The farther northwest one travels, the older the islands are. The apparent rate of movement away from the Hawaiian hot spot (deep source of magma) is about 9 to 10 centimeters per year. (Data from D. A. Clague and G. B. Dalrymple, U.S. Geological Survey Professional Paper 1350, 1987, p. 16.)

Earthquakes—close companions of active volcanoes—also show an unusual pattern in Hawaii. Although earthquakes occur along the entire length of subduction zones and oceanic ridges, in Hawaii they are common only at the southeast end of the chain, mostly beneath the Big Island.

The data from more than 30,000 microearthquakes occurring near Kilauea Volcano on Hawaii, as well as the pattern of tilting that indicates swelling and contraction of a shallow magma chamber, suggest an interesting subsurface model of a Hawaiian volcano (Figure 71).

Charting the location of the earthquakes at depth indicates the conduits through which the molten rock forces its way upward. The propelling force is the buoyancy of the lighter molten rock surrounded by denser crystalline rocks. As the magma rises—probably in many separate cracks rather than a continuous conduit—the hard surrounding

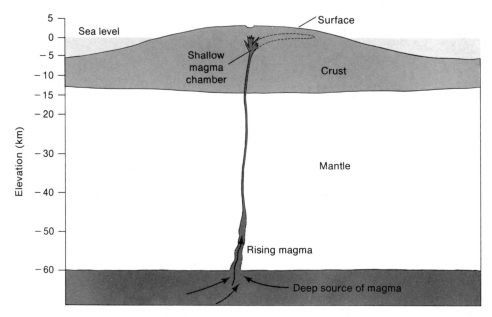

71 A schematic cross section of an active Hawaiian volcano. The vertical scale
is exaggerated two times. The dotted line extending to the right from the shallow
magma chamber is an inferred zone of feeder dikes to a rift zone. The deep source
of magma, below 50 to 60 kilometers, and the shallow storage zone at 3 to 6
kilometers beneath the surface, are the vital parts of the active system. (After
Jerry Eaton, U.S. Geological Survey.)

rocks are broken open by changing pressures and temperatures, causing earthquakes.

The deep source of magma appears to be within the plastic layer
50 to 60 kilometers beneath the Pacific Plate, the maximum depth of
the earthquakes. Diagrams of the locations of the intermediate-to-deep
earthquakes, 10 to 60 kilometers beneath Kilauea, have outlined over a
period of years a zone shaped like an inverted funnel 5 to 50 kilometers
in diameter and 50 kilometers high, which allows the deep magma to
rise into a shallow magma chamber 3 to 6 kilometers beneath the summit. Even though this shallow chamber is bounded by a region of many
microearthquakes, there are almost no earthquakes inside the chamber. Most earthquakes are thought to originate from the cracking and

slipping of brittle rocks that suddenly fracture. Plastic or liquid rock in a magma chamber can deform without breaking and thus produce no earthquakes (Figure 72).

The reason that magma chambers often form at these shallow depths is explained by the concept of neutral buoyancy, an idea championed by Mike Ryan of the U.S. Geological Survey. Below 6 kilometers the surrounding rocks are more dense than magma, thus causing it to rise; above 3 kilometers the surrounding rocks are less dense than magma, impeding its rise.

The tilt pattern near the summit of Kilauea supports this magma chamber model. Just as the pattern of tilting on the surface of an inflating balloon is related to the radius of the balloon, the pattern of tilt at Kilauea points to a pressure center about 4 kilometers deep. Continuing this balloon analogy, the deep source of magma is the pump; the inverted-funnel conduits are the connecting hoses; and the shallow magma chamber is the balloon.

The pattern of Kilauea's slow outward tilting (inflation) before an eruption and rapid inward tilting (deflation) during a flank eruption indicates a slow, almost continuous movement of magma from the deep source to the shallow chamber, and rapid and intermittent eruptions from the shallow chamber to the surface. Because the rate of movement of magma from depth to the shallow chamber often does not keep pace with the surface eruption rate, many eruptions stop until the shallow chamber has been recharged. The eruption at Puu Oo has been an exception. Since the steady-state lava emission began there in 1986, most of the lava has been slowly but continuously resupplied from depth. There has been a net deflation of Kilauea's summit during the fourteen years of the current eruption. But the calculated volume of magma that would cause this deflation is small compared with the total volume of lava that has erupted.

How do these processes relate to the origin and continuing vigor of the Hawaiian volcanic chain? There must be something unique about the location of the island of Hawaii. Somehow, deep magma must be more available at this location. Tuzo Wilson, a Canadian geophysicist, suggested an elegantly simple explanation that fits neatly with the movement of the Pacific Plate. He envisioned an upwelling of hot plastic rock originating deep beneath the plate. This upwelling, or *plume*,

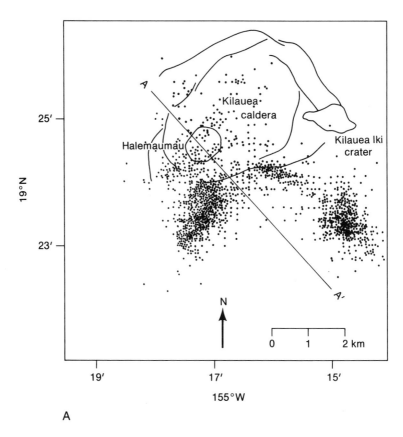

A

72 A. A map view of the locations of shallow microearthquakes underneath Kilauea Volcano in Hawaii. Earthquakes are caused by rock fracture, so zones of no earthquakes are regions of low stress or very low strength. Since high stresses are common near most active volcanoes, the lack of earthquakes in a region surrounded by them suggests a plastic zone occupied by magma. The dense clusters of microearthquakes shown on the map indicate the beginning of the rift zones. B. Cross section of Kilauea Caldera to a depth of 12 kilometers. (Data from Robert Koyanagi, U.S. Geological Survey.)

rises by very slow convection and causes some melting because of decompression as the hot rock rises in the mantle. The plume and its molten top, or *hot spot*, stays in the same location for millions of years as the plate slides over this continuing source of magma. A volcano grows over the hot spot and drifts away with the slowly moving plate, to

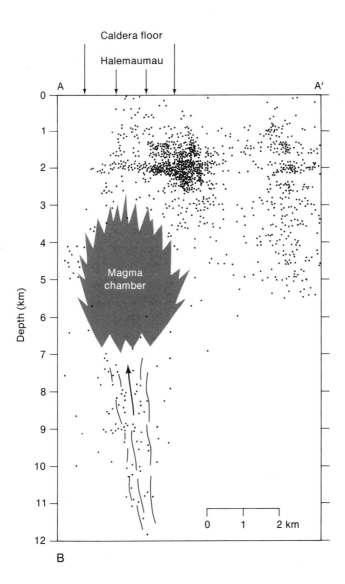

be replaced by a new volcano growing over the same hot spot, like smoke signals drifting in a gentle wind (Figure 73). Most geologists now accept the hot-spot idea, but the number of hot spots worldwide is still a matter of controversy.

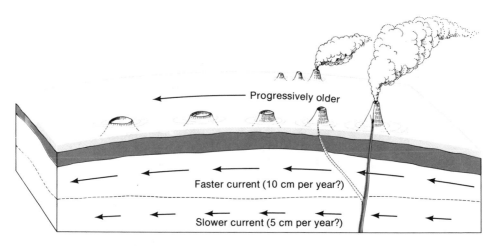

73   The original diagram used by Tuzo Wilson to suggest the origin of the Hawaiian Islands. The plate slowly moves over a relatively fixed source of magma, forming a new volcano as the older volcanoes are carried downstream. (From J. Tuzo Wilson, "Continental Drift." Copyright © 1963 by Scientific American, Inc. All rights reserved.)

The problem, of course, is explaining the character of a thermal plume. What keeps it going for millions of years, generating new magma at a relatively fixed spot in the Earth? The best idea now seems to be that the heat source of the hot spot is in the deep mantle, perhaps even at the boundary between the core and the mantle. This implies that the magma supplying hot-spot volcanoes has an even deeper source than does magma for the volcanic belts along the plate margins (Figure 74). The Hawaiian hot spot seems to be wide enough to feed three volcanoes at the same time—Mauna Loa, Kilauea, and the submarine volcano Loihi, off the south coast of the Big Island. It also appears that the Hawaiian hot spot has been generating magma for at least 80 million years, in order to build the Hawaiian Ridge and Emperor Seamounts (Figure 75).

If the hot-spot theory is correct, the aging of volcanoes away from the hot spot should match the rate of plate movement. The magnetic stripe pattern on the ocean floor indicates that the Pacific Plate is moving northwest at about 10 centimeters per year. If Kauai has drifted northwest during the 5 million to 6 million years since its lavas erupted, it should be 500 to 600 kilometers away from where it first formed, which is the present location of the Big Island. In fact, the actual distance is 560 kilometers, a striking verification of Wilson's hot-spot idea.

Beyond Midway Island, the Hawaiian chain is a submerged line of seamounts continuing northwest for another thousand kilometers. After

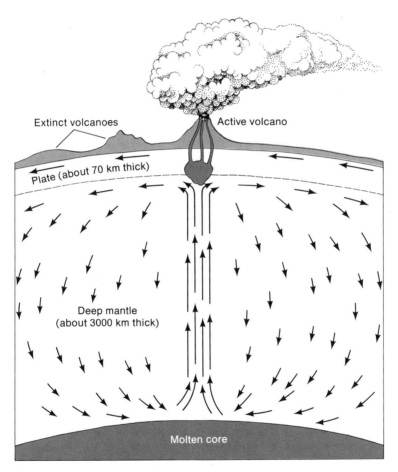

74   Deep mantle plumes generated by slow convection currents are one explanation of the relatively fixed position of hot-spot volcanic sources. This concept, originated by Jason Morgan, is illustrated here. (From G. B. Dalrymple, E. A. Silver, and E. D. Jackson, " Origin of the Hawaiian Islands," *American Scientist*, March 1973, p. 306.)

75 The world's current hot spots and their relation to the major plates. Hot-spot tracks, as illustrated by the Hawaii–Emperor trend, reveal past plate motions. Some geologists prefer a smaller number of hot spots. (After Kevin C. Burke and J. Tuzo Wilson, "Hot Spots on the Earth's Surface." Copyright © 1976 by Scientific American, Inc. All rights reserved.)

that, it meets a submerged mountain chain, the Emperor Seamounts, which trends north–northwest another 2500 kilometers toward Kamchatka. The Emperor Seamounts are thought to be a dogleg of the Hawaiian chain formed when the drift of the Pacific Plate was more northward than its present northwestward movement. The change in direction of plate movement, as estimated from the location of the bend, occurred about 43 million years ago.

The hot-spot explanation of the Hawaiian chain suggests that future islands will form to the southeast of the Big Island. The Loihi seamount, 28 kilometers southeast of the island of Hawaii, looks as if it will be that next island. Loihi is 970 meters below sea level, with two deep craters indenting its summit. Divers in submersibles have discovered pillow-lava flows on Loihi that appear to be very young. Warm springs with dissolved carbon dioxide have also been observed and sampled there.

Swarms of small earthquakes underneath Loihi were recorded in 1971–1972 and 1975, and they may have been related to submarine eruptions, though this correlation is by no means certain. During the summer of 1996, the most energetic earthquake swarm from Loihi ever recorded took place. Over a period of twenty days, the Hawaiian Volcano Observatory recorded more than 4000 Loihi earthquakes; 95 had magnitudes greater than 4, and the largest quake was a magnitude 5. Submariners at the University of Hawaii had been waiting for this kind of activity at Loihi, and dives were made in the submersible *Pisces* as the earthquake swarm waned. A new collapse crater was observed near the summit of Loihi, indicating that magma may have drained away from beneath that area, but no red lava or active eruptive vents were seen. The divers also encountered waters filled with fine white "dust" that reduced visibility to less than 2 meters. Whether vigorously active or not, it will probably be tens of thousands of years before Loihi becomes the next Hawaiian island.

The great strength of the plate-tectonics and hot-spot concepts is that their predictions are substantiated by clues found in so many places around the world. The plates have left their tracks in the form of volcanic scars and magnetic fields, and those tracks reveal a complex yet consistent global pattern.

At least two seamount chains in the South Pacific indicate other major hot spots beneath the Pacific Plate. Yellowstone Park, on the North America Plate, has been suggested as an active continental hot spot, with the Snake River volcanic plain its earlier track. The Galapagos Islands on the equator off the west coast of South America is another group of very active volcanoes of probable hot-spot origin.

The hot-spot idea even offers a solution to the Iceland puzzle: the problem of why Iceland is the only major part of the Mid-Atlantic Ridge out of water. Iceland is apparently a hot spot that straddles the ridge and has volcanoes of dual origin—both hot spot and mid-ocean ridge. This is an attractive explanation of the extra volcanism that has built Iceland above the sea.

No islands of similar origin could be more unlike than Hawaii and Iceland. Iceland is a rugged, nearly treeless land of harsh climate and stark beauty, inhabited by strong people of Viking heritage who have stalwartly maintained their culture and independence. Hawaii is a tropical garden, soft and seductive, inhabited by tourists who have almost submerged its Polynesian heritage. But both islands owe their distinctive beauty, power, and mystery to their origins in volcanic fire.

Sources

Anderson, D. L., T. Tanimoto, and Y. Zhang. "Plate Tectonics and Hotspots: The Third Dimension." *Science*, June 19, 1992.

Burke, Kevin, and J. Tuzo Wilson. "Hot Spots on the Earth's Surface." *Scientific American*, August 1976.

Clague, David, and C. Brent Dalrymple. "The Hawaiian-Emperor Volcanic Chain." In *Volcanism in Hawaii*. U.S. Geological Survey Professional Paper 1350, 1987.

Vink, Gregory, Jason Morgan, and Peter Vogt. "The Earth's Hot Spots." *Scientific American*, April 1985.

9

Lava, Ash, and Bombs

76   Pahoehoe lava flow, Hawaii. (Photograph by U.S. Geological Survey.)

A blast of burning sand pours out in whirling clouds.
Conspiring in their power, the rushing vapors
Carry up mountain blocks, black ash, and dazzling fire.

—Lucullus (first century A.D.)

Volcanoes are dark windows to the Earth's interior. Because their products are the only direct samples of the composition of the Earth's deeper levels, the aspect of volcanology that studies these materials has received considerable attention. Besides helping determine what the Earth is made of, the character of successive layers of volcanic deposits can reveal the type and sequence of prehistoric eruptions and provide a broader picture of volcanic events than present activity can.

Although most people think that lava flows are the main products poured forth from volcanoes, actually volcanic ash and larger solid fragments—called *volcanic cinders*, *bombs*, and *blocks*—are the major products of observed eruptions. Lumping together all sizes of solid fragments, geologists use the term *pyroclastic deposits* for this volcanic debris (pyroclastic literally means "fire fragment"). Pyroclastic materials are derived from three sources: magma that is cooled and broken into fragments by expanding gases at the time of eruption; fragments of old crater walls or magma conduits that are ripped loose in explosive eruptions; and clots of liquid lava thrown into the air that cool and solidify during their flight.

Pyroclastic materials are classified by the general size of the fragments. *Volcanic ash* can be as fine as flour but usually is more gritty, with particles up to the size of rice; *volcanic cinders* (also named *lapilli*—"little stones" in Italian) are larger than particles of ash and can be as big as golf balls; and blocks can be anything larger than cinders, including chunks the size of a house. *Volcanic bomb* is a special term for a cinder- or block-size clot of liquid lava thrown from an erupting vent (Figure 77). The brilliant arcs on time-lapse photographs of volcanic eruptions are the traces of volcanic bombs in flight.

77 A large volcanic bomb from Mauna Kea Volcano, Hawaii. This 50-kilogram specimen shows the twisted flow lines often acquired while these globs of hot plastic lava are spinning in flight. The pencil at the bottom shows the scale.

Pyroclastic materials are deposited in two different ways—by pyroclastic fall or by pyroclastic flow. Explosive volcanic eruptions often hurl and lift fragments to great heights, and as the particles fall back to Earth, they form a distinct layer or layers that blanket the slopes of the land. An Indian legend about a pre-1980 eruption of Mount St. Helens refers to a pyroclastic fall of volcanic ash as "gray snow that didn't melt." The falling process allows for some sorting of the debris: the coarser fragments fall first and nearby while the fine ash is winnowed away to fall last, sometimes at great distances. The deposit from a pyroclastic fall can be called just that, but for brevity some geologists use the term *airfall deposit*. An airfall deposit can usually be identified by

78   Volcanic ash layers in a road cut on Oshima Volcano in Japan. Airfall material is characterized by distinct layers that follow the slopes of the ground surface on which they fell. Two generations of ashfalls are seen here, separated by an erosional surface. The beds are not folded; the inclined layers are the original attitude in which they accumulated.

the way it blankets the topography and by the way its particles are sorted (Figure 78).

Sometimes an explosive eruption produces a cloud of volcanic debris so full of fragments that it is too heavy to rise very high into the atmosphere. Such an emulsion of gas and fragments forms a pyroclastic flow, the most dangerous kind of volcanic hazard. Pyroclastic flows can travel at speeds of more than 100 kilometers per hour, flattening and burning almost everything in their paths. Small pyroclastic flows often race down the valleys on a volcano's flanks, but larger masses that are expelled at high speeds or those accelerated by falling can sweep over small hills or across large flat areas in their path. Pyroclastic flow deposits often pond in low-lying areas after they lose speed. Large, thick pyroclastic flow deposits are distinct from airfall deposits because they exhibit only vague layering and almost no sorting of the finer and coarser fragments. Road cuts in airfall deposits show distinct bands of coarse and fine layers, often of different colors, while cuts in thick pyroclastic flows show massive deposits that look like pink or buff concrete.

The nomenclature of *pyroclastic flows* (the process) and *pyroclastic deposits* (the products) is both complex and confusing, because the descriptive terms come from several languages and the flows are difficult to observe. Pyroclastic flows can be deadly to close witnesses or hidden to more distant witnesses by the clouds of volcanic ash that shroud them. The challenge is to understand the nature of the pyroclastic flows from their deposits, which often leads to conflicting opinions. Here we condense and paraphrase the various types of pyroclastic flows and deposits described by Peter Francis in his excellent book *Volcanoes: A Planetary Perspective*.

Pumice flows deposit *ignimbrites* (Latin for "fire cloud rocks"), which are rocks or unconsolidated deposits of various-size fragments of pumice. *Nuées ardentes* (French for "glowing cloud") deposit blocks and ash composed of fragments more dense than pumice. *Pyroclastic surges* are flows of gas and fragments less dense than either pumice flows or nuées ardentes, and they leave thin but extensive deposits with cross-bedded layering (Figure 79). The general terms for all these flows and deposits are *pyroclastic flows* and *pyroclastic deposits*. Examples help distinguish these terms, and Francis provides many.

Huge pumice flows can form by the collapse or fallback of erupting clouds so laden with fragments that the convective lift from their internal heat can no longer sustain their rise. The 7700-year-old ignimbrite that spewed from around the stump of Mount Mazama as the caldera of Crater Lake was forming is an example of a massive, column-collapse pyroclastic flow. The volume of the Mazama ignimbrite is tens of cubic kilometers, and it extends as far as 50 kilometers around Crater

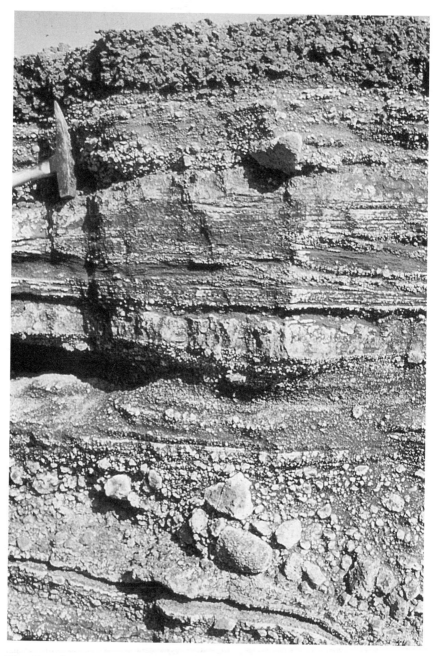

79 Pyroclastic surge deposits from the 1790 explosive eruption of Kilauea Volcano in Hawaii. The cross-bedding shows that wind velocities in the surges rose to speeds high enough to erode layers of previously deposited pyroclastic fragments. Transport direction was from left to right, and hammer head (upper left) indicates scale. (Photo by Robert L. Christiansen, U.S. Geological Survey.)

Lake. Some thick ignimbrites that were very hot when deposited consolidated into welded *tuffs*—hard, dense rocks containing pumice lumps that collapsed into small, flat obsidian disks.

Base surges are expanding rings of turbulent mixtures of fragments and gas that surge outward at the base of explosion columns. The base surge that formed at the nuclear test on the Pacific island of Bikini in 1946 rolled outward beneath the mushroom cloud for a distance of 4 kilometers at an initial speed of 50 meters per second. That was the speed of the ring's expansion; the turbulent internal wind speeds were probably much higher. The fluid dynamics of a base surge appear to be related to the formation and radial growth of a smoke ring.

Volcanic eruptions involving the explosive interaction of magma and water seem to be the ones that generate base surges, which are now called *pyroclastic surges* when related to volcanic explosions. Surge deposits are generally small in volume, extensive in area, and cross bedded like sand dunes and do not extend more than a few kilometers from their vents. The 1965 eruption of Taal Volcano in the Philippines, studied by Jim Moore of the U.S. Geological Survey, is a good example of pyroclastic surges and their deposits. The eruption took place near the shore of Volcano Island, a 5-kilometer-wide volcanic island in Lake Taal. During a two-day period, many horizontal blasts swept across the lake from the vent area. Trees near the vent were uprooted; farther away, the bark and wood on the sides of trees facing the vent were eroded away, but on their lee sides they were unaffected. Damage extended as far as 6 kilometers from the vent. More than 4 meters of dune-shaped surge deposits accumulated within a kilometer of the vent, but their thickness diminished rapidly to less than 1 centimeter 8 kilometers away. Mud was plastered on tree trunks and other vertical surfaces in the more distant parts of the devastated area, indicating that the surge at those locations was wet and below boiling temperature (Figure 80).

Nuée ardente (glowing cloud) was the name given by the French geologist Alfred Lacroix to the pyroclastic flow from Mont Pelée that annihilated St. Pierre. The flow was caused by the explosive breakup of a growing lava dome at the summit of the volcano, which then swept down on the city (Figure 81). One of the few witnesses to survive, Purser Thompson, on the steamship *Roraima*, anchored in the harbor, described the eruption:

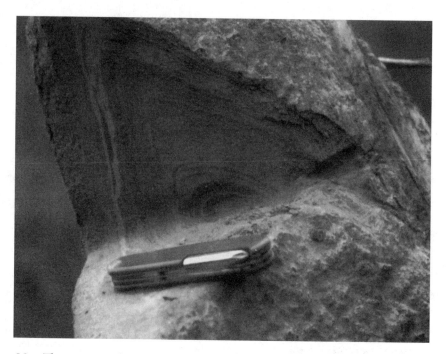

80   The eruption of Taal Volcano in the Philippines in 1965 generated pyroclastic surges that swept 2 to 3 kilometers across the southern part of Volcano Island. The sides of the trees facing the vent were plastered with layers of mud from the surges. The temperature of the hot surges that deposited these mud layers must have been lower than 100°C for the wet mud to stick to the trees.

For hours before we entered the roadstead, we could see flames and smoke rising from Mont Pelée. . . . There was a tremendous explosion about 7:45 o'clock, soon after we got in. The mountain was blown to pieces. . . . The side of the volcano was ripped out, and there hurled straight toward us a solid wall of flame. It sounded like thousands of cannon. The wave of fire was on us and over us like a lightning flash. It was like a hurricane of fire. . . . After the explosion, not one living being was seen on land. Only 25 of those on the Roraima, out of 68, were left after the first flash. ("The Destruction of the Roraima." *Frank Leslie's Popular Monthly*, July 1902.)

81   (*Facing page*) Large nuée ardente eruption from Mont Pelée, several months after a similar cloud of hot gases and volcanic ash destroyed St. Pierre and killed 28,000 people within minutes. (Photograph by A. Lacroix in 1902, courtesy of the American Museum of Natural History.)

Thought it must have seemed so, Mont Pelée was not blown to pieces. Although part of the old crater rim was destroyed, the flanks of the mountain were unchanged except for the rapid erosion and deposition in the valley that directed the glowing avalanche from the summit down to St. Pierre (Figure 82). The relatively small deposits from this pyroclastic flow consisted of dense blocks of volcanic rock from the growing lava dome, mixed with ash.

82   St. Pierre was totally destroyed by the nuée ardente (glowing cloud) that swept down from Mont Pelée on May 8, 1902. Only two people in the city survived. (Photograph by Underwood and Underwood from the Library of Congress.)

83   The cast of a dog buried by volcanic ashes at Pompeii. During the excavation, holes were found with skeletal remains inside. By carefully injecting the hole with plaster, a cast of the corpse is formed. About 2000 human victims have been unearthed at Pompeii. (Alinari/Editorial Photocolor Archives, Inc.)

A study of the explosive debris around a volcano tells us something about the nature of its previous eruptions and thus helps us predict the nature of possible future eruptions. For example, at Pompeii, the Roman city buried in A.D. 79 in pyroclastic deposits from Mount Vesuvius, the early stages of the eruption produced distinct airfall pumice deposits. Later accumulations of ash and pumice lumps indicate alternations of deposits from both pyroclastic surges and tephra. This sequence helps explain the relatively small number of bodies found buried at Pompeii in proportion to the size of the city. Although most people apparently escaped during the early rain of pumice fragments, stragglers were smothered and buried by the pyroclastic surges that followed (Figure 83).

Because many newspaper accounts of explosive volcanic eruptions still incorrectly refer to pyroclastic flows as lava flows, the general nature of each is probably more important than their details. A *pyroclastic flow* is a heavier-than-air mixture of hot gases and hot volcanic fragments whose loose packing and internal motions keep the mixture fluidized. It flows rapidly like a low-viscosity fluid, but after losing speed and fluidization, it dumps its fragments into pyroclastic deposits. A moving *lava flow*, in contrast, consists of viscous molten rock. It moves much more slowly than a pyroclastic flow, and as it cools and hardens it becomes solid rock.

Oceanic-ridge volcanoes and Hawaiian volcanoes erupt mainly lava flows. If we count the unseen flows deep beneath the sea, lava becomes the major product of all the Earth's volcanoes. Shallow submarine eruptions like Surtsey can build an island of explosive volcanic ash and bombs, but the great volume of magma from deeper eruptions goes into streams of underwater lava.

Individual lava flows are tonguelike in shape, much longer than they are wide. On the island of Hawaii, a typical lava flow might be 10 kilometers long, 200 meters wide, and 3 meters thick, but there is so much variation that any standard flow dimensions would be misleading. Each flow covers only a small fraction of the island with a finger of lava. This finger forms a low ridge, and later flows run either beside the ridge or between ridges formed by earlier eruptions. The process is like covering a jug with candle drippings; it takes hundreds of wax flows to build up a covering layer. The volcanic pile above sea level on the island of Hawaii is the accumulation of literally hundreds of thousands of lava flows.

A lava flow is hypnotic to watch. There is often a central river of orange red molten rock 5 to 10 meters across, flowing at speeds of 5 to 50 kilometers per hour, depending on the slope. The flow then oozes out on all sides from this central stream and forms slowly advancing dark lobes of cooling lava rubble riding on a molten but unseen core of the flow. This rubble of lava blocks tumbles down the steep front of the advancing flow, sometimes showing a glimpse of the glowing interior, and is slowly overridden by the advancing flow. The growing edges and fronts of the flow look like giant slow-motion bulldozer treads moving out, down, and under as the mass of the flow spreads forward. The perimeter of the dark, growing flow is much larger than the central river of glowing, fast-flowing lava (Figure 84).

In Hawaii, the kind of lava flow just described is called an *aa flow*, pronounced "ah ah." Its major characteristic is the rubble of broken lava blocks on its surface. Adventurers with thick boots have walked on

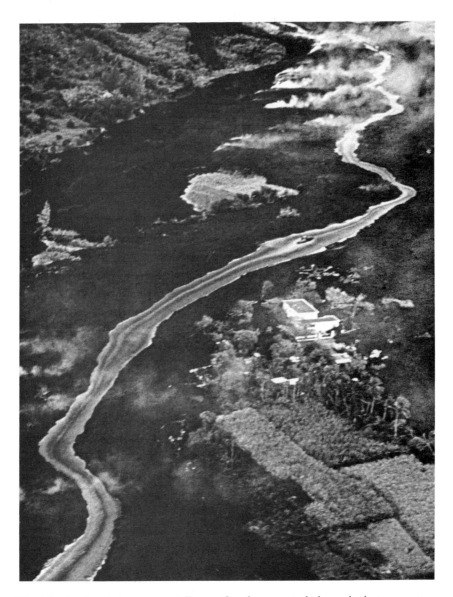

84   Flowing basalt lavas, especially aa, often have central channels that move at several kilometers per hour. These channels are red hot (in this photograph, light gray), in contrast to the partly cooled, black margins of the flow that advance more slowly, at rates of several meters per hour. (Photograph of the 1977 eruption of Piton de la Fournaise on Reunion Island by Pierre Vincent.)

these rubble surfaces while the flows are slowly moving, but aa flows are hard on boots even after they have stopped and cooled.

Another kind of Hawaiian lava flow is called *pahoehoe*, pronounced "pa hoy hoy." These flows are generally thinner than the aa flows and form a smooth-to-wrinkled surface of solid rock (Figure 76). Old Hawaiian foot trails favor the ancient pahoehoe flows, and for good reason. Bare feet will last for many more kilometers on their smoother, more solid surfaces (Table 1 and Figure 85).

Sometimes the main current within a pahoehoe flow crusts over and forms a tunnel filled with a fast-moving stream of lava. As the eruption wanes, the lava in this tunnel drains out, leaving an empty cave known as a *lava tube* inside the cooled flow. These tubes are 1 to 10 meters in diameter, and some can be followed underground for many kilometers.

TABLE 1
Forms of volcanic products

Form	Name	Characteristics (dimensions)
Gas	Fume	
Liquid	*Lavas*	
	Aa	Rough, blocky surface
	Pahoehoe	Smooth to ropy surface
Solid	*Airfall Fragments*	
	Dust	<⅛ mm
	Ash	⅛–2 mm
	Cinders	2–64 mm
	Blocks	>64 mm solid
	Bombs	>64 mm plastic
	Pyroclastic flows	Flows fluidized by hot gases
	Mudflows	Flows fluidized by rainfall, melting ice and snow, or ejected crater lakes
	Avalanches	Thick, hummocky deposits of broken rock

85   Lava flows of aa (left) and pahoehoe both issued from Kilauea Volcano in Hawaii in 1973. The dark, rough aa is about 3 to 4 meters thick; it was covering the smoother, glistening pahoehoe when the flow stopped.

The barren, smooth surfaces of pahoehoe flows, compared with the rubble that forms aa flows, are so distinctive that the Icelanders also have two separate words—*helluhraun* for pahoehoe and *apalhraun* for aa—to describe them where they occur in Iceland. The largest lava flow in recorded history was an aa (apalhraun) flow that issued from a 25-kilometer-long fissure in Iceland between June and November 1783. Named the Laki flow, it covered more than 500 square kilometers with lava, completely filling two deep river valleys in the process. The volume of lava produced by the Laki fissure during the several months of eruption was 12 cubic kilometers, enough to fill Yosemite Valley to a depth of 300 meters (Figure 86).

The texture of lava and pyroclastic rocks is largely controlled by the number and size of the gas bubble holes they contain. *Pumice*, being mostly holes, is one extreme. This frozen glass froth is so light that pieces

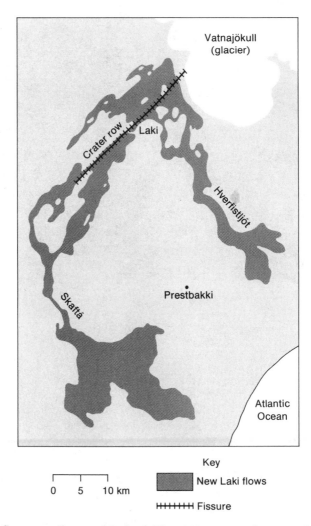

86   Laki flow in south central Iceland. The 1783 eruption from a 25-kilometer-long fissure produced 565 square kilometers (12 cubic kilometers) of basaltic lava flows, a world record. (Map data from Sigurdur Thorarinsson.)

of it float on water. Dense, solid lava without holes is the other extreme, with most volcanic rocks being somewhere in between (Figure 87).

Why do some volcanoes like Mount St. Helens explode, generating pyroclastic deposits, while others like Kilauea produce relatively quiet eruptions of lava? Several factors account for this difference, including gas content, viscosity, the number of sites where gas bubbles can grow, and the suddenness of the pressure release.

The gas content of magma is quite variable, ranging from less than 1 percent to more than 5 percent. At Mount St. Helens in 1980, the dissolved gases—mainly water and carbon dioxide—were estimated to be about 6 percent, by weight, of the magma; at Kilauea, the gas content of the magma amounts to about 1 percent. For this reason alone, Mount St. Helens had six times the explosive potential of Kilauea.

Mount St. Helens' magma was much more viscous than Kilauea's magma and confined the rapid gas escape in clots of magma until the pressures built to levels that explosively burst the clots into fragments. A firecracker analogy is useful here. Gunpowder confined in a paper wrapper explodes when ignited, while gunpowder poured from a broken firecracker into a small, open pile only burns. Of course, no combustion is involved in a volcanic explosion, but the nature of the gas release is analogous.

The places where bubbles in a liquid begin to form and grow are called *nucleation sites*. These sites in magma are often tiny crystals in the melt. The more nucleation sites that an erupting magma contains, the faster the gases can escape. Once small bubbles have formed, more gas will enter them, causing the bubbles to grow rapidly. Two analogies are useful to picture this process. Shake salt into a glass of beer and it will foam much more rapidly because you have added nucleation sites. Shake a can of soda pop before pulling the tab and it will be more explosive because you have shaken many small bubbles (nucleation sites) from the cap of carbon dioxide gas beneath the can lid into the liquid.

The suddenness with which the pressure surrounding a magma body is changed is also a very important factor. If magma rises slowly toward the surface (lower pressure) over months or years, the gases can escape slowly through cracks in the surrounding rocks. However, if the pressure changes suddenly, as it did at Mount St. Helens when the avalanche "uncorked" the shallow magma and superheated groundwater, the gases will expand explosively.

The 1980 eruption of Mount St. Helens had all four factors driving its violent explosion: the high gas content in the magma, the high viscosity of the magma, the many tiny crystals already present in the magma, and the sudden decompression caused by the avalanche sliding away. Kilauea, in contrast, has a low gas content and low-viscosity magma, fewer tiny crystals present as gas bubble nucleation sites, and a relatively slower release of pressure than at Mount St. Helens.

Before lava or pyroclastic fragments are erupted, their parent material is called *magma*. Magma is a melt of rock-forming elements often mixed with solid crystals and gas bubbles. The common elements in the melt are oxygen, silicon, aluminum, iron, calcium, magnesium, sodium, titanium, and potassium, all in varying amounts. It is common to report the amount of silicon and aluminum present in a melt in combination with oxygen. Hence silica, the compound of one silicon and two oxygen atoms (SiO_2), is the most abundant constituent; alumina (Al_2O_3) is second. The proportion of silica varies from about 45 to 78 percent of the total. Various types of magma and their resultant igneous rocks are classified by their silica content.

Geologists like to name rocks, and there are hundreds of varieties to identify. However, most volcanic rocks fall into a spectrum of compositions, with segments named basalt, andesite, dacite, or rhyolite, which have, respectively, a silica content of about 45 to 54 percent, 54 to 62 percent, 62 to 70 percent, and 70 to 78 percent. Rocks that contain more silica also contain more sodium and potassium and less iron, calcium, and magnesium. These last three elements, particularly the iron, form dark minerals, so that basalts are dark gray, almost black; andesites are medium gray; and dacites and rhyolites are usually light gray to tan. One important exception is obsidian, a dense, glassy form of rhyolite that was once used for arrowheads. Obsidian is nearly black.

87   (*Facing page*) Basaltic lava is often filled with gas bubble holes formed when steam and other gases dissolved in the magma are released by the low pressures at the Earth's surface. The tops of flows usually have more trapped bubble holes (vesicles) than do the centers of flows. The small crystals are olivine, one of the first minerals to form from the cooling melt. (Photograph at true scale by James Griggs, U.S. Geological Survey.)

Although there is less iron in rhyolite than in other kinds of volcanic rocks, in obsidian it is so finely dispersed as tiny crystals of opaque iron oxides that it makes the glass look shiny black (Table 2).

The chemistry of volcanic rocks is only part of the story of their complex composition. As magma cools, various minerals start to crystallize from the melt. Crystallization takes time. If basalt magma cools rapidly, within seconds or minutes, the compounds will not have time to arrange themselves into minerals, and the result is a dark, opaque glass. If the magma cools more slowly, over days or years, the minerals will have time to form and grow. Viscosity also affects crystal growth; crystals form more slowly in high-viscosity magmas, and so some thick rhyolite flows can be nearly all glass.

The cooling of molten basalt in Kilauea Iki lava lake in Hawaii is a good example of volcanic rock formation. The lava was erupted in 1959 at temperatures up to 1200°C and was nearly all liquid except for a small percentage of olivine crystals. Most of the water and other gases dissolved in the magma were vaporized and lost in the fume from the fire fountains as the lava sprayed out of the vent and then ponded in an older crater. The surface of the lake cooled quickly and formed a glassy crust with many bubble holes.

A project was set up to drill a sequence of holes in the thickening lava crust, to study the cooling and crystallization of basalt under the slowly changing and controlled conditions within the lava lake. While drilling, large volumes of water were pumped down the holes to prevent overheating the drill bits and core collectors, and temperatures were measured by electrical thermocouples lowered down the empty holes

TABLE 2
Types of igneous rock

Range of SiO$_2$ content	Extrusive (erupted on surface)	Intrusive (solidified below surface)
45–54%	Basalt	Gabbro
54–62%	Andesite	Diorite
62–70%	Dacite	Granodiorite
70–78%	Rhyolite	Granite

after the drilling had stopped and the holes had been reheated by the molten rock below.

The deeper crust formed more slowly because of the insulating effect of the overlying crust, and crystals of various minerals began to appear. Because the olivine crystallized first, at temperatures from 1250° to 1190°C, some crystals were present at the time of eruption. Pyroxene formed next, at temperatures between 1190° and 1180°C, and then plagioclase feldspar at 1170° to 1160°C. By the time the temperature had dropped to 1065°C, the basalt was half liquid and half crystals of plagioclase, pyroxene, and olivine, and it had become hard enough that steel probes could not be pushed into this mixture at the bottom of shallow drill holes. Magnetite began to crystallize at 1030°C, forming black mineral grains that give the basalt much of its dark color. By 980°C the basalt was all crystals except for 5 to 10 percent glass. Although it was still red hot, it had effectively become a solid.

During cooling and crystallization, the composition of each mineral is quite different from the overall composition of the original magma. This differentiation keeps changing the composition of the remaining melt fraction as the crystallization progresses (Figure 88). The most evident effect is in the composition of the final interstitial glass by the time cooling has reached 980°C. This glass has a silica content of 60 percent, compared with 48 percent in the original melt.

The cooling of andesite and rhyolite magma is equally complex, and other silicate minerals, including quartz, become important constituents of their crystallization into volcanic rocks. An important difference between rocks formed from a cooling silicate magma and ice formed by cooling water is the mixture of minerals in igneous rocks. Water freezes at 0°C and forms only one mineral, but magma freezes from 1250° down to 700°C, and the final, solid rock contains two or three or more minerals.

The volumes of lava and pyroclastic rocks produced in individual recorded eruptions range from a few cubic meters up to 50 cubic kilometers (Table 3). On average, subduction-zone volcanoes produce about 1 cubic kilometer of new volcanic rock each year, composed mostly of pyroclastic deposits.

The volume of volcanic rocks produced by oceanic-ridge volcanoes is largely hidden in the deep oceans, but it can be estimated indirectly. The spreading rates and sizes of the plates are reasonably well established,

TABLE 3
Volcanic Explosivity Index (VEI)

	0	1	2	3	4	5	6	7	8
General Description	Non-explosive	Small	Moderate	Moderate Large	Large	Very Large			
Volume of Tephra (m³)	1×10^4	1×10^6	1×10^7	1×10^8	1×10^9	1×10^{10}	1×10^{11}	1×10^{12}	
Cloud Column Height (km) Above crater		<0.1	0.1–1	1–5					
Above sea level				3–15	10–25	> 25 —————————————→			
Qualitative Description	"Gentle"	"Effusive"	←— "Explosive" —→	←—"Cataclysmic"	"Paroxysmal"	"Colossal"			
			←———— "Severe"	"Violent"	"Terrific"————→				
Eruption Type		←Strombolian→		←—————— Plinian ——————→					
	←— Hawaiian —→		←—Vulcanian —→←— Ultra-Plinian ————→						
Duration (continuous blast)	←———— <1 hr> ————→←——— >12 hrs ———→								
		←——— 1–12 hrs ———→							
CAVW[1] Max Explosivity	Lava flows← —————— Explosion or *nuée ardente* —————→								
		←— Phreatic —→ - - - - - - - →							
Tropospheric Injection	Negligible	Minor	Moderate	Substantial —————————————→					
Stratospheric Injection	None	None	None	Possible	Definite	Significant —————→			
Eruptions[2]	699	845	3477	869	278	84	39	4	0

[1] CAVW refers to Catalog of Active Volcanoes of the World, International Association of Volcanology and Chemistry of the Earth's Interior, Parts I–XXII, Rome, Italy, 1951–1975.

[2] Number of eruptions of each magnitude in Tom Simkin and Lee Siebert, *Volcanoes of the World,* Smithsonian Institution Press, 1994.

Source: This table adapted from C. G. Newhall and S. Self, The Volcanic Explosivity Index (VEI), *Journal of Geophysical Research 87,* 1231–38, 1982.

88   A photomicrograph of a very thin slice of Hawaiian basalt from Kilauea Iki lava lake. Drilling core holes into the slowly cooling lake provides samples of "quick-frozen" basalt in which the drilling water suddenly chills the slowly cooling melt and arrests the process of crystallization. The large crystal on the right is olivine, and the smaller, sticklike crystals are feldspar. The uniform gray area surrounding the silicate crystals is glass formed from the sudden cooling of the melt. (Photograph by the U.S. Geological Survey.)

and these data allow an estimate of the new area of seafloor that must be created every year to fill in the cracks opening between the spreading plates. The figure is about 2.5 square kilometers per year. Allowing 1.5 kilometers for the general thickness of basalt flows that accumulate on the seafloor, the average amount of oceanic-ridge eruptive products is about 4 cubic kilometers per year.

Hawaiian volcanoes average about 0.1 cubic kilometer per year of basalt lava flows, but the total for all hot-spot volcanoes is difficult to estimate because of the long intervals between eruptions in places like Yellowstone. A reasonable guess is an average of 0.5 cubic kilometer per year for the volcanic products of hot spots. The Earth's present volcanoes thus have a total output of about 5 to 6 cubic kilometers of new rock each year on the average. Many years go by without a large volcanic eruption, and then a Krakatau-type eruption comes along to catch up on the lag. In addition to the erupted rocks, shallow injections of magma intruded and crystallized beneath volcanoes also amount to large volumes, perhaps much more than the erupted rocks.

Huge deposits of pyroclastic flows that cover thousands of square kilometers and are tens to hundreds of meters thick are found in Japan, New Zealand, Central America, the western United States, and many other volcanic regions of the world. Some of these deposits give every indication that they were poured out in single, enormous eruptions that would dwarf Krakatau and Katmai. The volume of these deposits is on the order of 300 to 3000 cubic kilometers, compared with the 30 cubic kilometers of Katmai. This raises some fundamental questions. Was prehistoric volcanism, say a million years ago, greater than it is now? Or is the recorded history of volcanic eruptions so short—2000 years in Europe and about 200 years in the United States—that the data we have cover too brief a time to be representative? We think the latter is true. Krakatau and Katmai are probably only small samples of what nature can deliver in the way of volcanic cataclysms.

Sources

Cas, R. A. F., and J. V. Wright. *Volcanic Successions*. London: Allen & Unwin, 1987.

Fisher, R. V., and H.-U. Schmincke. *Pyroclastic Rocks*. Berlin: Springer-Verlag, 1984.

Francis, Peter. *Volcanoes: A Planetary Perspective*. New York: Oxford University Press, 1993.

Hibbard, M. J. *Petrography to Petrogenesis*. Englewood Cliffs, NJ: Prentice-Hall, 1995.

Simkin, Tom. *Terrestrial Volcanism in Space and Time*. Palo Alto, CA: Annual Reviews of Earth and Planetary Sciences, 1993.

Williams, Howel, and A. R. McBirney. *Volcanology*. San Francisco: Freeman, Cooper, 1979.

Lahars and Avalanches

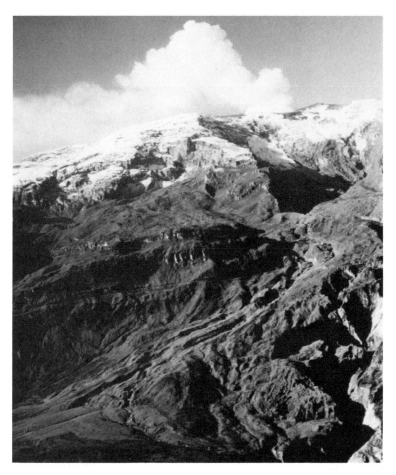

89 The summit of Colombia's Nevado del Ruiz Volcano, 5389 meters high, is covered by a 17-square-kilometer ice cap. A small pyroclastic eruption on November 13, 1985, melted about 5 to 10 percent of the ice cover and caused mudflows in the valleys draining the mountain's steep flanks. (Photograph by Robert Tilling, U.S. Geological Survey.)

Muddy: not clear, containing sediment, obscure in meaning.
                                               —Webster's Dictionary

I n stream valleys near explosive volcanoes, it is common to see terraces and banks composed of poorly sorted gravel in a matrix of sand, silt, and clay. These are deposits from *lahars*, the Indonesian name for volcanic debris flows—more commonly known as *mudflows*. Lahars can occur when heavy rains wash down loose ash from earlier eruptions or when an earthquake dislodges a steep slope of clay and rocks that have been deeply weathered by volcanic gas vents and hot springs. At other times and places, lahars can result when an eruption ejects a crater lake or rapidly melts the snow and ice on a volcano's high summit. As these huge floods of water surge down a volcano's flanks, they pick up volcanic ash and soil and quickly thicken into lahars. The surging mud in turn picks up larger rocks and boulders, and the entire mass thunders down a valley like a torrent of wet concrete.

The 1985 tragedy at Nevado del Ruiz, a 5389-meter-high volcano in the Andes Mountains of Colombia, was entirely the result of a lahar. Although Ruiz is only 5 degrees north of the equator, its high summit is covered with a large snow and ice field (Figure 89). In 1845 an eruption or earthquake triggered a lahar from Ruiz that killed about 1000 people, and the new town of Armero was built on top of it. In November 1985 a relatively small eruption of hot pumice and ash jetted from the crater, landed on the ice cap, and caused sudden melting. Huge floods of water mixed with ash poured off the summit, most of it rushing down the tributaries of the Lagunillas River, picking up soil, rocks, and trees as it swept along. In places these floods—now debris flows—scoured the canyon walls to heights of 80 meters above the normal stream levels.

The lahar roared down the Lagunillas Valley at speeds of 50 to 60 kilometers per hour. The town of Armero was located where the narrow canyon meets the wider, open valley of the Magdelena River. At the canyon's mouth the torrents spilled out on Armero in a debris flow as much as 30 meters high, which quickly spread out into dense flows 3 to 5 meters deep. Armero was swept away and more than 20,000 people died. The deposit of dried mud and debris left behind is about 1 meter thick and covers an area of 40 square kilometers. Boulders as large as 10 meters lie scattered where the town once stood (Figure 90).

When we wrote the first edition of this book, it was thought that the character of a mudflow deposit provided little or no information about the type of event that caused it. But now the eruption of Mount St. Helens (Chapter 5) and the studies of Kevin Scott of the U.S. Geological Survey and his colleagues have changed all that. As in most detailed studies, it's important to get the words straight before coming to conclusions.

Scott distinguishes two types of lahars: *cohesive debris flows* containing more than 3 to 5 percent clay (particles smaller than 4 microns)

90   The town of Armero, Colombia, 50 kilometers east and 5 kilometers below the summit of Ruiz Volcano, was destroyed by a massive mudflow that swept down the Lagunillas River from the November 13, 1985, eruption. About 22,000 people in Armero were killed by the sudden torrents of muddy water 3 to 5 meters deep.

and *noncohesive debris flows* containing smaller amounts of clay. A cohesive lahar maintains its "wet concrete" character throughout its course, but a noncohesive flow begins as a flood, thickens into a hyperconcentrated flow (one with 20 to 60 percent sediment by volume), continues thickening into a true lahar at more than 60 percent sediment, and then reverses this sequence as it loses velocity. The higher clay content in a cohesive lahar apparently holds the flow intact, in contrast to the noncohesive lahar, which picks up or loses its sediment load depending on its velocity. The character of the deposits from these two types of lahars thereby provides some clues to their origin.

For example, the large lahar that swept down the North Toutle River during and right after the May 18, 1980, eruption of Mount St. Helens was the cohesive variety. Its water and mud were derived in part from compaction and melting ice in the huge debris avalanche, and possibly from the giant waves of water displaced by part of the avalanche that plunged into Spirit Lake. The lahar from Ruiz that buried Armero was the noncohesive variety. The enormous, 3-cubic-kilometer Osceola Mudflow from Mount Rainier that reached Puget Sound 5700 years ago was a cohesive lahar generated by a major summit collapse and debris avalanche (Figure 91).

Although huge cohesive lahars occur less frequently than do the smaller, noncohesive lahars, they can travel long distances from their sources. A cohesive lahar can be triggered by a major earthquake as well as by a volcanic eruption.

A *glacial burst* is a unique type of volcanic flood that sometimes occurs in Iceland. A volcanic eruption beneath a thick ice cap can melt a huge volume of ice, and the meltwater accumulates in a lake or subglacial reservoir that is trapped under the ice. Increasing pressure from the reservoir's rising level eventually forces the water to move beneath the ice cap and escape through one of the river valleys that drains the glacier. These huge floods are called *jokulhlaups*—Icelandic for glacial bursts.

The most recent jokulhlaup occurred on November 5, 1996, five weeks after an eruption near Grimsvotn Volcano, a caldera under the Vatnajokull ice cap. The ice-covered lake that accumulated in the caldera was estimated to contain some 3 to 4 cubic kilometers of meltwater before it forced its way beneath the glacier. Grimsvotn Caldera is nearly buried by the ice cap, but some of its rocky rim protrudes through

91   Prehistoric mudflows from Mount Rainier, an active Cascade volcano, have covered part of what is now the Seattle–Tacoma Metropolitan Area. The giant Oceola Mudflow swept down the White River about 5700 years ago, covering more than 500 square kilometers. At nearly the same time, the Paradise Lahar rushed down the Nisqually River. As recently as 500 to 600 years ago, the Electron mudflow covered much of the lowlands along the Puyallup River. (After U.S. Geodynamics Committee, *Mount Rainier: Active Cascade Volcano*, National Academy Press, 1994, p. 25.)

the ice—similar to an atoll rising above the ocean. A seismograph that had been placed on this rim provided the first warning of the flood. As the water surged beneath the ice, it generated seismic vibrations that started ten hours before the flood burst out. This warning enabled the authorities to close the road that crosses the plain between the ice cap and the sea, and no one was killed. The peak discharge of the flood was

45,000 cubic meters per second, nearly three times the average flow volume of the Mississippi River. Ten kilometers of the road and one bridge disappeared, while two other bridges and 10 more kilometers of the road were badly damaged. Giant blocks of ice covered the plain as the flood subsided.

An *avalanche* is a large mass of snow, ice, soil, or rock—or a mixture of these materials—that falls or slides rapidly down a cliff or steep slope. An avalanche composed of a mixture of materials is called a *debris avalanche*. Large avalanches from the summit and steep slopes of volcanoes pose major hazards that may or may not be related to an eruption. A debris avalanche can be triggered by heavy rainfall or a strong earthquake without a volcanic eruption. However, at Mount St. Helens the great avalanche and the eruption were closely connected. The injection of a shallow body of magma caused the bulge on the already steep north slope of the volcano. This oversteepened slope was shaken loose by a moderate earthquake into the great avalanche, which in turn ripped open the mountain and unleashed the explosive eruption. These close connections between large avalanches and eruptions may be more common than previously thought.

Horseshoe-shaped craters that formed where a combination of avalanche and explosion has torn out a whole section of a steep volcanic cone are now being recognized on several active volcanoes. With knowledge gained from the Mount St. Helens avalanche and eruption, geologists have finally solved the mystery of the extensive area of hills and mounds of broken volcanic rock northwest of Mount Shasta in California. They now believe that between 300,000 and 380,000 years ago a gigantic avalanche from an ancestral peak of Mount Shasta rushed down the Shasta River valley for 43 kilometers. The hummocky debris-avalanche deposit covers 450 square kilometers and has a volume of 45 cubic kilometers. It is the largest on-land landslide known to have occurred anywhere in the world during the last million years (Figure 92).

Because of the steep slopes and high snow-and-ice-covered summits of stratovolcanoes, avalanches and lahars are more common on stratovolcanoes than on the more gently sloping and rounded shield volcanoes. But such assumptions are sometimes shaken by new discoveries. For instance, undersea surveys around the shield volcanoes of the Hawaiian Islands during the past decade have revealed some submarine landslides that make the Shasta avalanche look small.

92   This landscape of hills north of California's Mount Shasta (in the background) is part of a gigantic avalanche deposit that slid from the volcano sometime between 300,000 and 380,000 years ago. This debris avalanche traveled more than 40 kilometers beyond the base of the mountain and contains 45 cubic kilometers of broken volocanic rock. (Photograph by C. Dan Miller, U.S. Geological Survey.)

In 1983 the United States declared sovereignty over seabed resources extending for 200 nautical miles off its shores, and the U.S. Geological Survey was given the task of mapping and evaluating this new area. One of the most useful tools in this survey was side-scanning sonar towed by surface vessels that could image a swath of sea bottom as much as 50 kilometers wide. The seabed off the shores of the Hawaiian Islands from Midway Island to the island of Hawaii was mapped between 1986 and 1991. The astounding outcome of this survey was the discovery of at least sixty-eight giant underwater landslides more than 20 kilometers long. Streams and waves had been thought to be the great destroyers of the Hawaiian volcanoes—instead it turns out to be gravity.

Two dominant types of Hawaiian submarine landslides have been identified—slumps and debris avalanches. A *slump* moves on an average slope of more than 3 degrees and involves a large mass of a volcano that slides with little internal disruption. A 100-kilometer-wide block can extend from an on-land volcanic rift zone to its submarine terminus more than 20 kilometers at sea and be as much as 10 kilometers thick (Figure 93). According to studies of the southeast flank of Kilauea Volcano, the movement of a slump block is apparently either a slow creep of about 10 centimeters per year or a sudden slip of a few meters that generates a major earthquake like the magnitude 7.2 quake of 1975.

In contrast, a giant submarine *debris avalanche* slides on an average slope of less than 3 degrees and can move slightly upslope past the axis of the Hawaiian Trough. It moves rapidly and breaks up some of the avalanche block into fragments. An avalanche is thinner and longer than a slump, and its run-out is marked by hummocky terrain with broken blocks as much as 1 to 10 kilometers in diameter. Some avalanches, such as Alika 1 and 2 off the west coast of the Big Island, have well-developed on-land escarpments where they slid away. These steep slopes have been draped with younger lava flows from Mauna Loa Volcano. The largest debris avalanches have volumes on the order of 5000 cubic kilometers—the largest avalanches known on Earth.

The Hilina slump on the southeast coast of the Big Island is still active, but the youngest avalanche (Alika 2) has an inferred age of about 100,000 years, based on tsunami deposits on the Hawaiian island of Lanai. Can a slump turn into an avalanche? Some apparently have,

93 (*Facing page*) Giant submarine landslides on the flanks of the Hawaiian Islands are among the largest on Earth. Some of them are longer than 200 kilometers, and some have volumes of thousands of cubic kilometers. Two types are seen on the sonar surveys: (1) Large slumps that move by creep and intermittent small displacements during large earthquakes. The Hilina slump on the southeast flank of Kilauea Volcano is an active example. (2) Huge debris avalanches that move long distances at high speeds and that are caused by catastrophic failure. These avalanches, like Alika 2 on the west coast of Hawaii, are thought to have generated enormous tsunamis. Fortunately, their recurrence interval is in the range of hundreds of thousands of years. (After James Moore and William Normark, Palo Alto, CA: Annual Reviews of Earth and Planetary Science, 1994, p. 128.)

A

B

where the toe of a slump has become oversteepened. Could the active Hilina slump suddenly slide away? Perhaps. Geologists have speculated that a very large Hilina earthquake exceeding magnitude 8 might trigger a catastrophic failure of the southeast flank of Kilauea Volcano. *Perhaps* and *maybe* are favorite words of geologists; what is known is that two giant tsunamis left their marks on the islands of Lanai and Molokai 100,000 and 200,000 years ago, respectively, and these huge waves are thought to be related to the disturbance of the ocean by large submarine avalanches.

The Lanai tsunami washed blocks of coral into extensive surface deposits as high as 325 meters above current sea level, and the Molokai wave deposits are as high as 70 meters above current sea level. In 1975 a tsunami with a run-up height of a few meters washed up on the southeast coast of the Big Island soon after the magnitude 7.2 earthquake on the Hilina slump. Like many geologic processes, small events occur frequently and large events more rarely. Is a catastrophe with an average recurrence time of hundreds of thousands of years worth worrying about? For a person whose average life span is less than one-thousandth of this recurrence time, the answer is certainly no; there are far more dangerous things to worry about. What is fascinating is not the danger of these giant underwater landslides but the fact that they were virtually unknown only a few years ago.

Other surveys are already turning up similar landslides at several volcanic islands. These include Reunion Island in the Indian Ocean, Stromboli and Mount Etna in the Mediterranean, Tristan de Cunha and the Canary Islands in the Atlantic, and the Marquesas Islands in the Pacific. What started as a survey to evaluate the resources on the seabed near Hawaii revealed an entirely new stage in the life cycle of oceanic volcanoes.

Sources

Carlowicz, M. "Glacial Burst Overwhelms Skeidara." *EOS* (American Geophysical Union), November 19, 1996.

Crandell, D. R. "Gigantic Debris Avalanche of Pleistocene Age from Ancestral Mount Shasta Volcano, California, and Debris-Avalanche Hazard Zonation." *U.S. Geological Survey Bulletin* 1861, 1989.

Moore, James, W. B. Bryan, and K. R. Ludwig. "Chaotic Deposition by a Giant Wave, Molokai, Hawaii." *Geological Society of America Bulletin*, July 1994.

Moore, James, W. R. Normark, and R. T. Holcomb. *Giant Hawaiian Landslides*. Palo Alto, CA: Annual Reviews of Earth and Planetary Science, 1994.

Moore, James, W. R. Normark, and R. T. Holcomb. "Giant Hawaiian Underwater Landslides." *Science*, April 1, 1994.

Scott, Kevin. *Origins, Behavior, and Sedimentology of Lahars and Lahar-Runout Flows in the Toutle–Cowlitz River System*. U.S. Geological Survey Professional Paper 1447-A, 1988.

Scott, K. M., J. W. Vallance, and P. T. Pringle. *Sedimentology, Behavior, and Hazards of Debris Flows at Mount Rainier, Washington*. U.S. Geological Survey Professional Paper 1547, 1995.

Cones and Craters

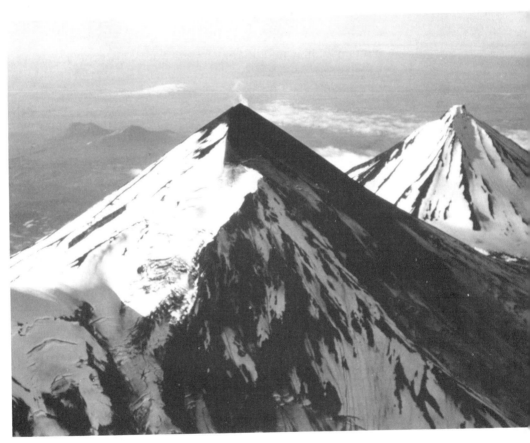

94 Pavlof and Pavlof's Sister, stratovolcanoes in the Aleutian Islands, Alaska. (Photograph by Katia Krafft.)

163

Only to a magician is the world forever fluid, infinitely mutable
and eternally new. Only he knows the secret of change, only he knows truly
that all things are crouched in eagerness to become something else,
and it is from this universal tension that he draws his power.

—Peter Beagle (1976)

Mountains come in different styles. Some, like the Himalayas, were thrust upward where the crust of the continents is telescoped together by powerful compressive forces. Others, like the ranges between the Rockies and the Sierra Nevada, were formed by regional uplift and the downfaulting of keystone blocks into the basins between the ranges as the crust pulls apart under tension. Still others, such as the Appalachians, resulted from regional uplift and the subsequent etching by erosion of hard and soft rock formations—folded by earlier compression—to form ridges and valleys.

Volcanoes have a style all their own. They are built upon the landscape by outpourings of new rock. Their forms change rapidly, some even during a person's lifetime. In 1975 in the Tolbachik region of Kamchatka, a new cinder cone grew to a height of 600 meters in just four months. The known record for rapid growth goes to Vulcan, a volcano on the rim of Rabaul Caldera in Papua New Guinea. The 1937 eruption lasted only four days, and most of Vulcan's 230-meter-high cone was built during the first twelve hours.

Geologists learn that the hills are not everlasting, that erosion slowly strips away the mountains and fills the intervening basins with their debris. It is usually taken for granted these days that river valleys were carved out by the streams they contain. But move to an active volcanic terrain and all the rules change. The mountains are growing, new rocks are created, and the valleys are often the spaces between lava flows. In other words, some valleys are formed not by erosion but by lack of construction. Even experienced geologists must remind themselves that they are in a strange land where the rules are often reversed.

Volcanoes create some of the most beautiful and unusual landscapes on Earth. Their power and fire command respect and are reminders of the intense vitality of our Earth. *Kazan*, the Japanese word for volcano, means "fire mountain"; our word *volcano* comes from Vulcano, the name of an Italian island whose volcanic peak was considered the forge of Vulcan, the Roman god of weapons.

The shapes of volcanic mountains range from the perfect cones of Mount Fuji in Japan and Mayon in the Philippines to flat lava plateaus in Iceland that barely meet the designation of mountain. The volcanic vents on these mountains range from vast circular basins called *calderas*, many kilometers in diameter, to tiny craters a few meters across (Figure 95). Several factors are at work in producing these great variations in form.

The shape of the vent from which the volcanic products escape to the surface is of prime importance. Often the vent is a long, nearly vertical crack in the ground, several hundreds to thousands of meters long and deep, and only a few meters wide. Magma rises through this fracture and erupts along a linear vent at the surface, sometimes welling out in

95   Summit of Cotopaxi Volcano, Ecuador. The nested craters resulted from explosive eruptions of different intensities. Snow and ice dominate the summit of this 5900-meter peak despite its volcanic activity and its location less than 1 degree south of the equator. (Photograph by G. E. Lewis, U.S. Geological Survey.)

wide streams of lava and sometimes in a line of lava fountains known as a *curtain of fire*. The Laki eruption in Iceland and the beginning of the Puu Oo eruption in Hawaii are examples of eruptions from linear vents.

A specific location along a linear vent usually becomes the principal source of volcanic emissions as an eruption progresses, forming a central vent from which a volcanic mountain begins to grow. If a volcano is eroded down to its roots, the hardened magma in the exposed feeding cracks can be seen as long, low ridges with greater resistance to erosion. These features are called *volcanic dikes*. The eroded remnant of a steep, pipelike conduit that fed a central vent is called a *volcanic neck* (Figure 96).

96   Ship Rock in New Mexico is a volcanic neck with radiating dikes. Erosion has stripped away the volcanic cone, leaving only the hard skeletal remains. Ship Rock is 450 meters high, and the dikes form great walls stretching across the desert. (Photograph from John S. Shelton, *Geology Illustrated*, p. 15. W. H. Freeman and Company, Copyright © 1966.)

97   The lava dome that forms Lassen Peak in California was erupted about 27,000 years ago. The 2-cubic-kilometer volume of the dome puts it among the world's largest. Lassen last erupted between 1914 and 1917, with the principal explosive activity occurring in May 1915.

The viscosity of the magma as it erupts also has a profound impact on the shape of a volcano. Very viscous lava squeezes up into a steep-sided plug over its vent, called a *lava dome*. Lassen Peak, a 27,000 year-old volcano in northern California, is one of the world's largest known lava domes (Figure 97). Solid fragments thrown from a vent form a pile of debris called a *cinder cone* around or downwind from the crater (Figure 49). These cones have straight sides with slopes of about 30 degrees, the angle of repose above which the cinders slide until a stable angle is established. Very fluid lavas, on the other hand, flow long distances on gentle slopes, forming lava plateaus or low-sloping volcanic piles called *shield volcanoes*. Since the composition of a lava is generally related to its viscosity—basalt is more fluid and rhyolite more viscous—the shape of a volcano is an important clue to its composition (Figure 98).

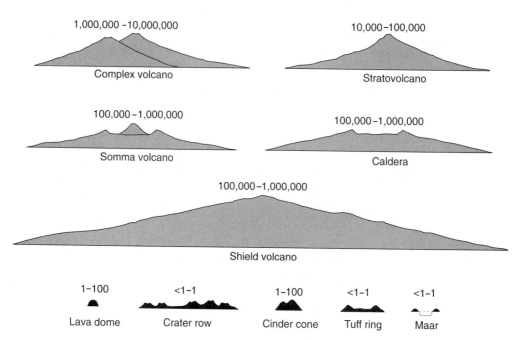

98 The relative sizes, shapes, and life spans of various types of volcanoes. The shaded profiles are vertically exaggerated two times, and the black profiles, four times. Their approximate life spans of active volcanism are indicated in years. (Adapted from Tom Simkin and Lee Siebert, *Volcanoes of the World,* Smithsonian Institution Press, 1994.)

Many *subduction volcanoes* begin erupting with an explosion of volcanic ash that is followed by flows of lava. The alternation of ash and flows forms the steep concave slopes of the classic volcanic cone. The scientific name for this type of structure is *stratovolcano*, or *composite cone*, because of the layering of pyroclastic deposits and lava flows (Figures 94, 99, 100, and 101).

The surface environment of the volcanic vent also affects the shape of the resulting volcano. Submarine volcanoes are the best examples of the effects of environmental factors. As explained in Chapter 3, deep submarine volcanoes are not explosive because the water pressure is so great that steam cannot form and expand. But because water cools the lava faster than air would, the lava piles are generally steeper than those same lavas would be if they erupted above water. Shallow submarine eruptions usually are explosive, even when they involve basalt magma that would be poured out on land in relatively quiet fountains and flows.

Tuff rings and maars are closely related volcanic landforms that are built by steam explosions from rising magma entering groundwater at

99 Initial stages of an explosion cloud forming on Lassen Peak in California. This lava-dome volcano erupted from 1914 to 1917. (Photograph by B. F. Loomis, June 14, 1914, courtesy of the National Park Service.)

100   *(Facing page)* Mayon Volcano, in the Philippines, erupting. The rising ash column and descending pyroclastic flow partly conceal the perfect symmetry of Mayon's stratovolcano cone. (Photograph by Dainty Studio, Daraga, Albay, Philippines, April 30, 1968.)

101   Santiago Crater of Masaya Volcano in Nicaragua reveals the inner structure of a stratovolcano. The light layers are lava flows, and the dark layers are ashfalls. The thick dark layer at the top is part of a cinder cone whose summit is just to the left of this photograph.

a shallow depth. A *maar* is a flat-floored explosion crater with a low rim of debris derived largely from the country rock of the excavated crater rather than new volcanic material. Its crater often fills with ground-water, forming a circular lake. The Eifel craters in Germany are examples of maars. A *tuff ring* is built by multiple pyroclastic surges that build a ring of debris around its explosion crater. The deposits that build the ring are a combination of the excavated country rock and new volcanic material. Diamond Head in Honolulu, Hawaii, is a classic example (Figure 102).

Volcanic eruptions beneath thick piles of glacial ice show many of the same features as submarine eruptions: deep outpourings of pillow lava, followed by shallow-water explosions of volcanic debris, and

102   Diamond Head, the prominent landmark east of Waikiki Beach in Honolulu, looks quite different when seen from the air. This tuff ring was built by a late phase of Hawaiian volcanism, some 3 million years after the main eruptions created the island of Oahu. Pieces of coral in the tephra indicate that the steam explosions generated by the interaction of erupting lava and seawater occurred in or below the shallow reefs that fringe the island.

103   Herdubreid, a table mountain in Iceland. Pillow lava on the steep flanks covered by gently sloping flows on the summit indicates that table mountains were formed by eruptions beneath a glacial ice cap. The elevation to the edge of the table marks the thickness of the ice. (Photograph by Gudmundur Sigvaldason.)

capped by gently sloping flows of lava that reached above the glacial surface. The table-mountain volcanoes of Iceland are thought to have originated in this way (Figure 103). The glacial ice caps are now largely gone, but the steep shoulders on table mountains mark their former thickness.

Other factors, such as the volume of erupted material and the length of time between eruptions, also influence the shape of a volcano. Small pyroclastic-flow deposits fill valleys, while huge pyroclastic-flow deposits create *ignimbrite plateaus* that bury all the underlying topography. Los Alamos, New Mexico, is located on such a volcanic plateau, formed about 1 million years ago (Figure 104). Ignimbrite plateaus are common in many volcanic regions of the world, but fortunately this type of eruption has not taken place in recorded history. The great size and speed of such a pyroclastic flow would be cataclysmic.

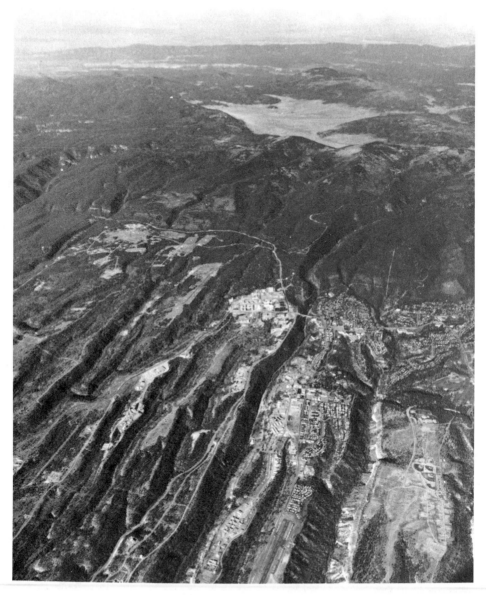

104   Los Alamos, New Mexico, the birthplace of nuclear power, lies on the flank of a giant prehistoric volcano. The city sprawls across a canyon-cut plateau of thick pyroclastic flow deposits, erupted about 1 million years ago. Ejection of this huge volume of material caused the volcanic summit to collapse, forming Valle Grande, the 20-kilometer-diameter caldera covered by grasslands in the background. (Photograph by William Regan of the Los Alamos Scientific Laboratory.)

The craters on volcanoes also are created in many different ways. Funnel-shaped craters usually result from an initial enlargement of the vent by volcanic explosions, followed by the collapse of loose debris back into the crater at the end of the eruption. After gas escapes from magma, the reduction in volume sometimes allows the magma to drain back down the vent, forming a deep cylindrical crater on the volcano's summit. The escape of magma from a vent on a volcano's flank can also cause subsidence of the summit crater. For example, the crater on Mount Fuji is about 700 meters in diameter and 100 meters deep, with nearly vertical walls.

When a volcano erupts a very large volume of magma, the void created underground cannot support itself, and so part of the volcano subsides, or collapses, into the emptied space. The collapse creates a large circular basin with steep walls, sometimes several hundred meters high; many such basins are the remains of the summit of an earlier volcanic cone. *Collapse calderas* form within a few hours or days, as at Krakatau, and are generally associated with great eruptions. Long Valley Caldera, near Mammoth Mountain, California, on the eastern side of the Sierra Nevada, extends 30 kilometers east–west and nearly 20 kilometers north–south. It apparently resulted from the collapse associated with the 700 cubic kilometers of pyroclastic flows that erupted violently and deposited the apron of ignimbrites that surround the caldera. Crater Lake in Oregon is another collapse caldera. It formed during a great eruption that occurred 7700 years ago, as dated by the carbon 14 in charcoal from trees that were incinerated beneath the pyroclastic flows (Color Plate 22 and Figure 105).

Calderas are prominent features of several Hawaiian volcanoes. The summit of the major shield volcano Mauna Loa is the rim of a large elliptical caldera 3 by 5 kilometers in diameter, with vertical walls as high as 200 meters (Figure 106). This caldera formed about 1200 years ago, and the absence of explosive volcanic debris around its rim indicates that a major removal of magma from its summit chamber, probably by large flank eruptions of lava, caused the caldera to collapse.

Craters and calderas are difficult to tell apart, and in fact, Kilauea Caldera is often called Kilauea Crater, even on official maps. A practical though arbitrary distinction can be made on the basis of size. Craters are smaller than 1 kilometer in diameter, and calderas are larger than 1 kilometer in diameter. Another distinction is that calderas are generally

South

Airfall pumice
and ash

North

The eruption began with explosive
jetting of ash and pumice fragments
into a high column.

2 km

2 km

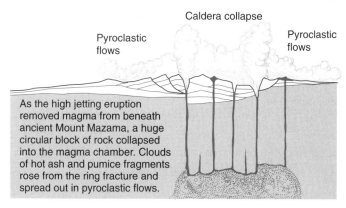

Caldera collapse

Pyroclastic
flows

Pyroclastic
flows

As the high jetting eruption
removed magma from beneath
ancient Mount Mazama, a huge
circular block of rock collapsed
into the magma chamber. Clouds
of hot ash and pumice fragments
rose from the ring fracture and
spread out in pyroclastic flows.

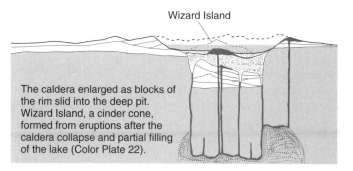

Wizard Island

The caldera enlarged as blocks of
the rim slid into the deep pit.
Wizard Island, a cinder cone,
formed from eruptions after the
caldera collapse and partial filling
of the lake (Color Plate 22).

105   Schematic cross sections showing the great eruption and caldera collapse
that formed Crater Lake, Oregon, about 7700 years ago. (Drawings by Richard
Hazlett, after Charles Bacon, U.S. Geological Survey.)

106 Circular craters and a large 3-by-5-kilometer caldera indent the snow-covered summit of Mauna Loa Volcano in Hawaii. These collapse features with cliffs up to 200 meters high lie at an elevation of about 4000 meters, at the top of a huge gently sloping shield volcano. Historic eruptions have been filling the craters and caldera with numerous flows. The view is to the north. (Photograph by the U.S. Army Air Force in 1939; courtesy of the National Archives.)

formed by collapse (Color Plate 22, Figure 105), while craters form by either collapse or explosion.

Craters are sometimes hidden beneath domes of viscous lava; Lassen Peak in California is an example. Rows of small craters along a fissure erupting lava can be buried by the thick flows that issue from them. The lack of cones and craters on extensive lava plateaus such as the Columbia River lava fields in Washington and Oregon makes it difficult to locate the actual vents from which the lavas erupted.

Although volcanic mountains evolve through time, their mature and old-age forms are far from simple. A cinder cone may be the early stage of a stratovolcano. Caldera collapse may then swallow up the summit of a large stratovolcano, and new lava domes may subsequently appear over vents along the rim of the caldera. One way out of this classification dilemma is to call the result a *complex volcano* (Figure 107). Still, with careful geologic mapping it often is possible to distinguish the component parts of a complex volcano and the order in which they formed. The history of eruptive habits and their sequence can be very useful in attempting to forecast the future hazards at a particular volcano.

Volcanoes can be erupting, dormant, or dead; their lifetimes are extremely variable. The great volcanic complex near Los Alamos, New Mexico, has been erupting intermittently for 15 million years. Most individual volcanoes in Iceland erupt only once, suggesting that Surtsey's lifetime was only three and a half years. The average life span for a recurrently erupting volcano is roughly 100,000 to 1 million years. In the end, erosion takes over and the stumps of vanquished volcanoes join the other passive mountains of the world.

The time of a volcano's death can be estimated by the degree of erosion, but such estimates must be made with caution. Erosion itself is extremely variable: fast in humid climates, slower in cold, dry regions, and fastest of all in cold, wet (glaciated) regions. Hawaii has large changes in climate within short distances. The low slopes facing the trade winds are hot and humid and receive more than 5 meters of rainfall per year. The summit of Mauna Loa is cold and relatively dry, and on the lee shore of the island of Hawaii, the climate is that of a desert. Lava flows only fifty years old have been reclaimed to soil and jungle in the hot, humid areas, but high on Mauna Loa, flows that are a thousand years old look freshly erupted. In the high deserts of Mexico, cinder cones remain almost untouched by erosion for thousands of years.

But climates change and the tooth of time finally wears down all volcanoes. Since its formation a million years ago, half the Los Alamos plateau has been incised by deep canyons. Islands like Hawaii erode

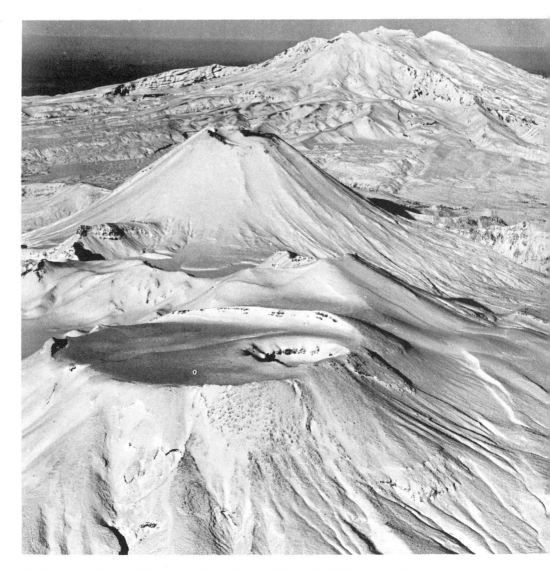

107 An aerial view of Tongariro, Ngauruhoe, and Ruapehu Volcanoes in New Zealand. The crater of Tongariro, in the foreground, is 1.3 kilometers wide and is transitional between a large crater and a small caldera. It is filled by ponded prehistoric flows cut by a small collapse crater. The cone of Ngauruhoe, in the middle ground, is typical of an active stratovolcano with a summit crater. Ruapehu, in the background, is also an active stratovolcano, but it contains a crater lake whose eruptions have prematurely eroded the volcano's flanks. (Photograph by S. N. Beatus, New Zealand Geological Survey.)

away in 5 million to 10 million years and then slowly subside beneath a growing coral atoll cap.

We have records of volcanic activity for 2000 years in Europe, 1000 years in Iceland, and only a little more than 100 years in the Cascade Range of the northwestern United States. Comparing this with a million-year volcanic life span and a similar time to erode away the dead cones and craters, it is folly to declare a volcano extinct just because it has had no eruptions in recorded history. Any volcanic peak that shows little of the ravages of time, such as Mount Rainier, Mount Hood, or Mount Shasta, is only sleeping.

Sources

Bullard, Fred. *Volcanoes of the Earth*. Austin: University of Texas Press, 1984.

Green, Jack, and N. M. Short, eds. *Volcanic Landforms and Surface Features*. New York: Springer-Verlag, 1971.

Krafft, Maurice. *Volcanoes: Fire from the Earth*. New York: Abrams, 1993.

Scarth, Alwyn. *Volcanoes*. College Station: Texas A&M University Press, 1994.

Wood, Charles, and Jurgen Kienle, eds. *Volcanoes of North America*. Cambridge: Cambridge University Press, 1990.

12

Roots of Volcanoes

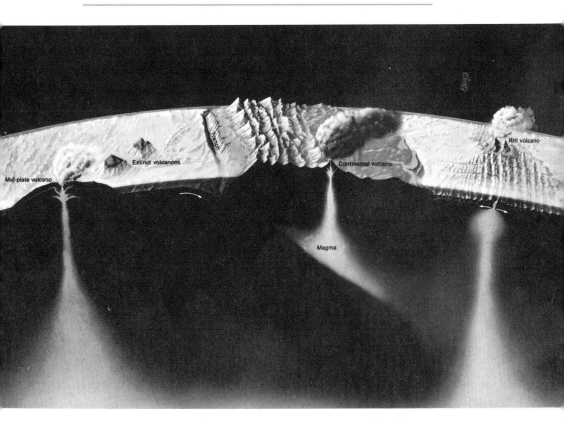

108   A schematic diagram of the three types of volcanoes related to plate tectonic processes. Right: Rift volcanoes formed by separating plates, such as the submarine volcanoes on the Mid-Atlantic Ridge and Icelandic volcanoes. Center: Volcanoes formed by converging plates at subduction zones, such as Japan and the Andes Mountains of South America. Left: Hot-spot volcanoes, such as those in Hawaii and the Hawaiian Ridge. (From *Powers of Nature*. Copyright © 1978, National Geographic Society. All rights reserved.)

Long is the way
And hard, that out of hell leads up to light.

—John Milton (1608–1674)

When John Wesley Powell was trying to establish the U.S. Geological Survey, a senator from Nebraska challenged him, saying: "You can't see any farther into the Earth than any other man." Even today geologists can hardly deny the truth of this statement, but we do have a great deal of indirect data on the inside of the Earth that helps guide our speculations.

For example, we can study the seismic waves from large earthquakes because they travel through the Earth to seismographs around the entire globe. Earthquake waves change in character as they travel away from their source, but regardless of the location of the earthquake, seismographic recordings made at equal distances from the source are nearly the same. This result can best be explained by an interior structure of the Earth that is symmetrical around the center (Figure 109). Thus if the Earth's interior is composed of various materials, they must be arranged in layers or spherical shells like an onion.

We also have clues to the weight of the Earth's interior. We know the average density of the whole Earth from its size and from the magnitude of the force of gravity at the surface: it amounts to 5.5 grams per cubic centimeter, which is twice the density of surface rocks. Therefore, some parts of the Earth's interior must be very dense material to bring the average up to 5.5.

Reasonable speculations about the roots of volcanoes must be consistent with our knowledge of the physics and chemistry of the whole Earth as well as our observations of the surface geology. One fact is quite clear: the surface geology indicates that our Earth is very complex and heterogeneous. The physical and chemical data on the Earth's interior yield average values rather than details, however, so most models of the Earth's interior show simple shells of rock or molten metal. As a first approximation these cartoons are probably accurate, but the de-

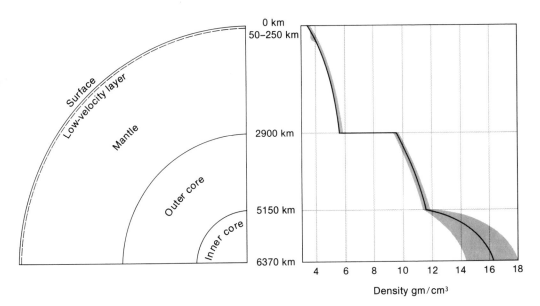

109   The estimated density of the interior of the Earth. The solid line in the graph on the right shows the most probable value of density at each depth, and the shaded region outlines the range of uncertainty. (Data from K. E. Bullen, "The Interior of the Earth." Copyright © 1955, by Scientific American, Inc. All rights reserved.)

tails are missing. Filling in this picture is the most fascinating challenge in geoscience research today.

These details are not trivial. Diamond and graphite are the same chemical, but wars have been fought over their difference. Synthetic diamonds can be made from graphite with complex high-pressure, high-temperature furnaces reaching ranges of 100,000 atmospheres (103,320 kilograms per square centimeter) at more than 1000°C. Natural diamonds occur in volcanic pipes that originated at depths as much as 200 kilometers beneath the surface; thus the diamonds, their inclusions, and the pieces of wall rock erupted with them provide a direct sample of the composition, pressure, and temperature at depths in the Earth far below the reach of drills (Figure 110).

The uplift and erosion of ancient volcanoes expose their shallow roots that were once as much as 10 kilometers below the surface. Post-mortems on these rocks show that the dikes and pipes of chilled

SOUTHERN AFRICA

110   A schematic cross section through southern Africa based on inclusions brought up in diamond pipes and other volcanic pipes in which diamonds are absent. Spinel and garnet are minerals characteristic of the depth ranges indicated. The boundary labeled G/D is the transition between graphite and diamonds. (Modified from Francis R. Boyd, Carnegie Institution, *Year Book 1986*, p. 98.)

magma often had connected surface vents to larger storage chambers of molten rock at depths of 2 to 10 kilometers beneath the surface (Figure 111). These shallow magma chambers were complex bodies of intersecting dikes and layers of molten rock or, sometimes, more cylindrical masses called *stocks* or *plutons*. Their composition, the texture of their crystals, and the host rocks at their edges yield important clues to how the magma chambers evolved and how and when they cooled into crystalline rocks. In general, the composition of shallow magma chambers is the same as that of surface volcanic products, although their texture is more coarsely crystalline because they cooled more slowly. For example, granite is the root rock in stocks beneath volcanoes that once erupted rhyolite lava flows or huge volumes of rhyolitic pyroclastic flows (Table 2).

Shallow magma chambers are not the deep roots of volcanoes, however, because they occur at depths too shallow for rocks to be melted by the Earth's heat. Uplift and erosion have not exposed the

111   A schematic diagram of the surface forms and subsurface structure of various volcanic features. (Adapted from R. G. Schmidt and H. R. Shaw, *Atlas of Volcanic Phenomena*, U.S. Geological Survey, 1972.)

TABLE 4
Radioactive heat production from uranium, thorium, and potassium in common igneous rocks

Rock	Amount of Radioactive Element in Rock (parts per million)			Amount of Heat Produced (joules/kilogram year)
	Uranium	Thorium	Potassium	
Granite	4	13	4	.03
Basalt	0.5	2	1.5	.005
Peridotite	0.02	0.06	0.02	.0001

deep roots of volcanoes; they are still hidden from direct observation, so our information about them is more tenuous. The sources of the heat that melts rocks was once a subject of lively debate. The leading contenders were the Earth's original heat from the time of formation; heat from the breakdown of the radioactive isotopes of uranium, thorium, and potassium; heat from the gravitative energy of redistributing heavy elements toward the center of the Earth; and heat from the tidal friction of slowing the Earth's rotation. Most geologists now agree that the decay of radioactive isotopes is the principal heat source.

The accretion of the planets at the time the solar system was formed generated immense amounts of heat from collision and gravitational collapse. Whether or not the planets actually melted depended on the speed of their accretion. They would have melted more completely if they had formed rapidly, because the heat would have had less time to be lost to space. Even if the original Earth were molten, the heat of its formation alone could not account for the estimated heat lost during its 4.5-billion-year life span.

Moreover, even if the Earth had formed cold, the radioactive decay of unstable atoms would have started a warming trend—warming fast at first because of many short-lived, highly radioactive elements, and then more slowly from the longer-lived but less radioactive elements, such as uranium 238, thorium 232, and potassium 40. The heat generated from the estimated remainder of these three isotopes in the Earth today can account for all of the world's currently escaping heat (Table 4).

As the primitive Earth was initially being heated by radioactive decay, melting and the gravitational segregation of iron to form the core would have greatly hastened the heating process, perhaps melting the entire Earth. Even so, this gravitational melting would have been a one-time process and its heat long gone unless sustained by other processes.

The heat produced by tidal friction gets its energy from the slowing down of the Earth's rotation. (Some studies of the growth bands of 400-million-year-old fossil corals indicate that the early Earth's rotation was faster and the days shorter.) Much of this energy is dissipated in the swirling tides of the Earth's oceans, but some geologists believe that the tidal warping of weak zones inside the solid Earth may be another continuing source of heat. As in many geologic debates, the problem is complicated by too many plausible explanations. Probably more than one of the various theoretical sources of energy are involved, and all of them are virtually inexhaustible in terms of human time spans.

Regardless of the exact source of the heat in the Earth, much is known about the variation in temperature near the surface. In deep mines and drill holes, the temperature increases with depth. This rate of increase, called the *thermal gradient*, averages about 30°C per kilometer of depth. By measuring the rocks' heat-insulating quality, which is quite high, the actual loss of heat from inside the Earth can be calculated. This quantity is about 5000 times less than the heat that reaches the Earth from the sun. Even so, it is an enormous amount of energy, much greater than the more spectacular heat loss from all the Earth's volcanoes.

At 30°C per kilometer, the 800°-to-1200°C melting temperatures of rock should be reached at depths of 30 to 40 kilometers beneath the Earth's surface. There is good evidence, however, that the thermal gradient is not constant, but rather that it slows at greater depths, because of the way that the radioactive isotopes of uranium, thorium, and potassium are distributed in the Earth. These elements have an affinity to granitic rocks and tend to be concentrated in the Earth's crust. The production of heat from their radioactive decay therefore diminishes with depth, reducing the rate at which temperatures increase with depth. The best estimates of this effect indicate that temperatures inside the Earth reach 800° to 1200°C at depths of 60 to 100 kilometers beneath the surface. This is the same depth as the top of the seismic low-velocity layer described in Chapter 1—an important piece of independent evidence supporting the temperature calculations. The roots of volcanoes must then extend down at least 60 to 100 kilometers beneath the surface to reach a source of primary magma (Figure 112).

Seismic evidence indicates that the low-velocity layer is not a zone of completely liquid melt. Rather, the rock is only partially molten; perhaps a small percentage of the total material is liquid and is contained in a spongelike mass of weak but solid rock at a high temperature. How then does the magma separate and rise toward the surface?

The density of magma is less than that of the rock from which it melts. Under the influence of gravity, the lighter molten material tends to rise, and the residual solid material to sink. Any fractures reaching deep into the Earth would hasten this upward movement of magma, which would tend to escape in the same manner as petroleum gushes

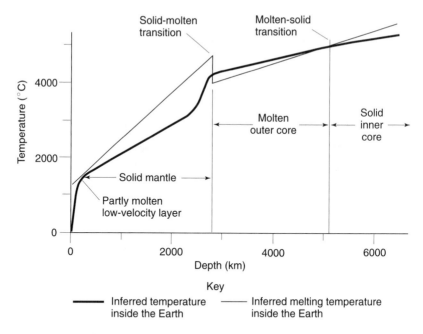

112   A graph of inferred temperature (black line) and inferred melting temperature (gray line) inside the Earth. There are four control points for this graph: (1) The thermal gradient near the Earth's surface is about a 30°C increase per kilometer of depth. (2) At a depth of about 100 kilometers, the actual temperature and melting temperature curves must almost touch to account for the low strength and partial melting of the low-velocity layer. (3) For the lower mantle to be solid and the outer core molten, the actual temperature and melting temperature curves must cross. (4) For the outer core to be molten and the inner core to be solid, the actual temperature and melting temperature curves must again cross. (Data from Reinhard Boehler, 1996.)

from a wild well. The problem is the conduit. Do fractures reach down through the rigid crust into the melt zone, or do bodies of magma push and melt their way slowly upward as great rising blobs to feed the shallow magma chambers? (Figure 113).

Separating plate margins have major fractures, and volcanoes along the mid-ocean ridges probably have conduits along these fractures between the plates. In these rift zones the thermal gradients are much higher than average, and seismic evidence indicates that the low-velocity layer comes up to within 10 or 20 kilometers of the surface.

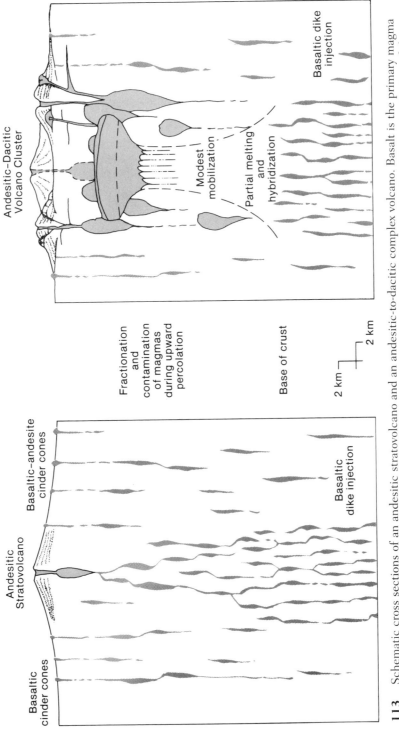

113 Schematic cross sections of an andesitic stratovolcano and an andesitic-to-dacitic complex volcano. Basalt is the primary magma rising from depth, but it becomes more silica rich by the removal of early-formed crystals (fractionation) and by contamination of the basalt with crustal rocks of higher silica content. Andesitic-to-dacitic magma chambers are shown by the dotted pattern. (From Wes Hildreth, *Journal of Geophysical Research* 86 [1981]: 10179.)

The following labels appear in the figure:

Andesitic Stratovolcano

Basaltic–andesite cinder cones

Basaltic cinder cones

Basaltic dike injection

Fractionation and contamination of magmas during upward percolation

Base of crust

2 km

2 km

Andesitic-Dacitic Volcano Cluster

Modest mobilization

Partial melting and hybridization

Basaltic dike injection

At converging plate margins, island-arc volcanoes occur in belts parallel to the ocean trenches. The volcanic belt is often, but not always, offset 100 to 200 kilometers onto the overriding plate (Figure 53). The generation of magma seems to be aided by water and carbon dioxide from the subducting slab, where it reaches a depth of 100 to 150 kilometers. This variation in offset argues for magma's forcing its way upward rather than rising along some special fracture zone.

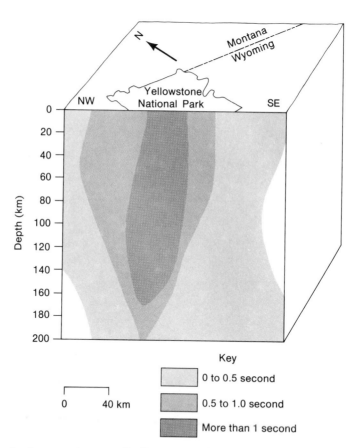

114 A schematic diagram of Yellowstone Park showing the seismic velocity anomaly at depth. Vibrations from distant earthquakes are delayed more than 1 second during their passage through the region shown. The carrot-shaped zone may include a huge volume of still-molten rock. (Data from H. M. Iyer, U.S. Geological Survey.)

In Hawaii the location of earthquakes below the shallow magma chamber of Kilauea indicates that there is a zone of fractures shaped like an inverted funnel about 5 to 50 kilometers in diameter and from 10 to 50 kilometers deep, reaching down to the top of the low-velocity layer. The shape of this region of active cracking does not imply a major fracture zone parallel to the Hawaiian Island chain, but rather a local zone of failure closely associated with the ascending magma. Here it appears that the magma forces its own way up through many small and disconnected fractures to the shallow magma chamber beneath Kilauea. The rate of inflation of the summit area during the years from 1960 to 1986 indicated that magma was being fed continuously from depth into the magma chamber at 3 to 6 kilometers below the surface, at rates of about 100,000 to 500,000 cubic meters per day. From 1986 through 1996, Puu Oo has been erupting in a nearly steady-state manner, and the summit of Kilauea has shown only a few brief periods of inflation.

No one has ever seen the deep roots of volcanoes, but improved geophysical techniques can provide some startling indirect views. In Yellowstone Park, small delays in the transit time of vibrations from distant earthquakes indicate that a large body of magma or very hot rock underlies much of the region. By recording these delays of only seconds at many seismograph locations, geophysicists can map the rough size and shape of the huge thermal anomaly.

Another seismic technique, a spin-off from prospecting for oil by the reflections of sound waves generated at the surface, extends the "X-ray vision" of geophysicists to a depth of 50 kilometers. Analysis of the deep seismic cross sections obtained by this technique reveals a great deal of complexity in the rock structures at depth. In the Rio Grande Rift section of New Mexico, a major sound-reflecting zone at about 20 kilometers beneath the surface is thought to be a layer of magma. An adaptation of this technique to seismic exploration beneath the seafloor reveals the top of a shallow magma chamber along the axis of the East Pacific Rise (Figure 115).

A new branch of geophysics called *seismic tomography* uses the travel times of earthquake waves that have passed deep within the Earth's interior. By comparing the small differences in actual travel times with those for a uniformly layered model of the Earth and integrating the data from hundreds of earthquakes and thousands of seis-

115 A seismic reflection profile of the crest of the East Pacific Rise near 14°14′ South. The reflection labeled 2A at a depth of about 200 meters below the seafloor beneath the ridge axis is interpreted to come from the base of the lava flows that overlie dikes. The strong reflection about 1 kilometer below the seafloor labeled AMC (axial magma chamber) is interpreted to come from the top of a thin layer of molten rock. The base of the oceanic crust, labeled MOHO, lies at a depth of about 7 to 8 kilometers below the seafloor. (From Robert Detrick et al., *Science* 259 [22 January 1993]: 501.)

mograms, geophysicists are able to draw three-dimensional plots of some of the larger thermal and compositional anomalies in the major layers of the Earth's interior. As is usually true, the closer you look at a feature, the more complex it turns out to be. The once simple circle that separated the Earth's core and mantle is now being redrawn into complex chemical and thermal boundary layers. For geologists and volcanologists who are getting their first geophysical look at magma and the deep roots of hot spots, these are exciting times.

Sources

Ahrens, Thomas, ed. *Global Earth Physics*. Washington, DC: American Geophysical Union, Reference Shelf no. 1, 1995.

Anderson, Don, T. Tanimoto, and Y. Zhang. "Plate Tectonic and Hotspots: The Third Dimension." *Science*, June 19, 1992.

Boehler, R. *Melting Temperature of Earth's Mantle and Core: Earth's Thermal Structure*. Palo Alto, CA: Annual Reviews of Earth and Planetary Sciences, 1996.

de Silva, S. L. "Antiplano–Puna Volcanic Complex of the Central Andes." *Geology* 17 (1989).

Smith, Robert, and R. L. Christiansen. "Yellowstone Park as a Window on the Earth's Interior." *Scientific American*, February 1980.

Wyllie, Peter (chair), and the Committee on Status and Research Objectives in the Solid-Earth Sciences. *Solid-Earth Sciences and Society*. Washington, DC: National Academy Press, 1993.

13

Origin of the Sea
and Air

116 Sea, steam, and flying rocks: Hawaiian lava flow enters the ocean. (Photograph by
G. A. Macdonald, U.S. Geological Survey.)

The smell of sulphur is strong, but not unpleasant for a sinner.

—Mark Twain (1866)

The brimstone smell of sulfur dioxide at active volcanoes is noxious, especially to people with respiratory problems, but usually not fatal. Another gas commonly exhaled at volcanic vents is more insidious, however. Carbon dioxide—a colorless, odorless gas—is heavier than air and can accumulate in concentrations that can suffocate animals and people with little or no warning. A tragic example of this occurred at Lake Nyos in Cameroon, West Africa, on August 21, 1986.

Lake Nyos fills a 1-to 2-kilometer-wide, 200-meter-deep crater on a dormant volcano that had not erupted in recorded history. Without warning a rumbling burst of gas, mostly carbon dioxide, belched from the lake and poured into the valley below. A survivor who lived above the lake heard the rumbling and in the darkening evening light saw a gas cloud that looked like a smoking river moving down into the valley. He felt a blast of wind from the lake and was gagged by its awful smell.

The dense cloud of gas—later estimated to be as much as 1 cubic kilometer in volume—flowed down the valley below the crater for 10 kilometers, killing 1700 people and 3000 head of cattle. Houses and farms were undamaged, but dead bodies were everywhere.

Scientists who investigated the disaster agree that carbon dioxide gas of volcanic origin was the killer, although they disagree about just what happened. Most of them think that for hundreds of years, magmatic gases had been seeping into the lake from below and, by the high pressure of the overlying water, were dissolved and concentrated in the deep water near the lake bottom. This rare phenomenon can occur in a tropical lake because the warm, less dense layer of water near the surface prevents the lake water from overturning, which would release most of the gases dissolved at depth. In temperate regions, cooling of the surface water in winter eliminates the density layers, and wind and currents can bring the deep water to the surface, so that any accumulated gas is released harmlessly.

Once the deep water in a tropical crater lake has become saturated with carbon dioxide, the potential for its rapid effervescence—like the sudden opening of a champagne bottle—exists.

Although most scientists agree up to this point, they disagree on what triggered the sudden release of dissolved gas in the lake in Cameroon. Some feel that large amounts of cool rainwater entering Lake Nyos or a landslide on the steep wall of the lake basin could have triggered the release of gas. Once the effervescence was started, it would stir the lake and continue the rapid de-gassing. Another school of thought is that a small volcanic eruption beneath the lake triggered the gas outburst, and a few investigators believe that all the carbon dioxide was of sudden eruptive origin. No new volcanic rock has been found around the lake, but evidence for this interpretation may lie hidden at the bottom of now-calm Lake Nyos.

But catastrophic events like this are rare. In a larger view, over geologic time volcanic gases have helped make the Earth—with its abundant surface water—a special planet in our solar system (Figure 116). Of all the inner planets—Mercury, Venus, Earth, and Mars—only the Earth has oceans. Venus has a massive, dense atmosphere of carbon dioxide and sulfuric acid that perpetually hides its rocky face in clouds, but its 500°C surface temperature is too hot to allow the condensation of an ocean.

What is the origin of the water and gases on the Earth, and why are the inner planets so diverse in this respect? Many factors seem to be involved: the planet's distance from the sun, the composition of the nebular gases from which it condensed, the planet's mass, the presence or absence of an original atmosphere, the changing temperature of its interior, and the amount of water and other gases bound up with the rocks and metals that formed the planet. The distance from the sun strongly affects a planet's surface temperature, and the planet's mass controls its gravity. Light gases escape into space when the surface is warm and the gravity low. For these reasons, the moon's surface is a vacuum: it does not have enough mass to hold an atmosphere at its distance from the sun.

The question of how much of the Earth's atmosphere was originally present—that is, formed at the time the planet formed—is the basis of a controversial and lively debate that also involves the role of volcanoes.

If the Earth formed slowly by accretion from a nebular dust cloud, it may have remained cool enough for an original atmosphere to form. By contrast, if the accretion was rapid and hot or if the surface temperatures were higher than at present because of either internal processes, such as radioactive heating, or external processes, such as excessive heat from the early ignition of the sun, any original atmosphere would have boiled away.

The "neon argument" supports the concept of the loss or lack of any original atmosphere. Neon is an inert gas; it does not combine with silicate minerals or metals to form solid compounds. Neon also is relatively abundant as a gas in the cosmos, and its density indicates that it is not subject to loss into space from the Earth's present atmosphere. An original atmosphere of the Earth that derived from a cosmic dust cloud should have contained a modest amount of neon. But the Earth's present atmosphere has only a tiny fraction of neon, much less than would be expected in an original atmosphere. Its near absence indicates that the original neon was lost along with all the other gases and water from the hot surface of a primitive Earth. A later atmosphere and oceans evolving from volcanic gases escaping to the Earth's surface would not include neon. Only gases such as water, carbon dioxide, and others that could be chemically bound into solid matter would be released by melting and eruption on the Earth's surface.

For these reasons, many geologists think that part of the Earth's oceans and atmosphere boiled out of the Earth's interior, particularly during the period of melting from the conversion of gravitational energy as its dense core formed. Alternatively, collisions with the abundant icy comets that are thought to have populated the early solar system may have been the major contributor of water and carbon dioxide to the young Earth's surface.

Before the concept of plate tectonics was firmly established, it was thought that volcanic gases were the main source of the Earth's oceans and atmosphere and that this was a long evolutionary process. Now this thinking has shifted, and volcanoes are viewed as recycling the oceans and atmosphere instead of creating them. The general argument goes like this: water, carbon dioxide (in calcium carbonate sediments), and other volatile compounds are carried down into the mantle on seafloor crust at subduction zones. After a few million years these volatiles are exhaled by eruption at subduction zone volcanoes or, after a few hundred million years, by eruption at rift volcanoes. Hot-spot volcanoes may also recycle some of the volatiles incorporated into the low-velocity layer at subduction zones, but the ratio of helium isotopes in hot-spot volcanic gases suggests that they have a deeper, more primitive source.

With this idea in mind, let's look at the actual data on gases escaping at active volcanoes. Molten basalt rising to the Earth's surface contains about 0.5 percent, by weight, dissolved gases. In sharp contrast, silica-rich magmas have a much more variable gas content and can contain as much as several percent gas, by weight. These gases are in solution in the same way that carbon dioxide gas is dissolved in beer. They escape by effervescence into gas bubbles and foam as the magma reaches the surface, just as beer foams when its containing pressure is suddenly released. The amount of gases dissolved in magma is the principal factor controlling the explosiveness of eruptions.

It is not easy to determine the composition of volcanic gases, since they escape during eruptions. Huge volumes of gases are involved in explosive eruptions, but it is nearly impossible to get close enough to sample them before they mix and react with unknown amounts of air (Figure 117). Indirect methods of sampling have proved more successful. Gases in deep submarine eruptions are frozen into the glassy rinds of pillow lava, and these trapped gases have been analyzed by several investigators and provide some of the best estimates of the composition of volcanic gases currently available.

Sometimes crystals forming in a cooling magma chamber enclose a tiny pocket of melt. If these crystals are erupted, the melt pocket quenches to an inclusion of glass in which the volcanic gases are trapped. An analysis of these inclusions by Fred Anderson at the University of Chicago revealed that magma from subduction volcanoes has a much higher gas content than does magma from submarine-rift volcanoes.

Richard Stoiber of Dartmouth College pioneered still other techniques for analyzing volcanic gases. He uses instruments called *absorption spectrometers* to measure the sulfur dioxide in volcanic fume clouds backlighted by the sky. Stoiber also analyzes gases adsorbed on volcanic ash particles collected as they fall from the eruption clouds. His studies indicate that each active volcano generates about as much sulfur gas

117   Volcanic gases escaping from this crack in a cinder cone formed during an eruption of Kilauea Volcano in Hawaii are depositing an encrustation of minerals. Wayne Ault of the Hawaiian Volcano Observatory measures the temperature of the vent and collects gas for chemical analysis. (Photograph by the U.S. Geological Survey.)

pollution as does a large coal-fired electric power station. Table 5 gives Stoiber's estimates of the annual emission of major gases from volcanoes on land.

A volcanic gas seldom makes headlines, but several newspaper articles have recently been published about the emission of carbon dioxide seeping up through the ground near Mammoth Mountain, a popular ski resort in eastern California. This seepage of gas was first noticed in 1990 when localities were found in which all the trees had been killed, and it was confirmed when a forest ranger almost died from suffocation inside a snow-survey cabin. Since 1980 the general area—a giant caldera that collapsed 700,000 years ago—has been undergoing earthquake swarms and uplift (Figure 118). The last eruption occurred here 500 years ago and was small compared with the giant eruption that created the caldera, but it definitely establishes that Long Valley Caldera is only dormant, not extinct. Geologists monitoring the earthquakes, uplift, and now emission of carbon dioxide are not forecasting an imminent eruption, but they do think that dikes of magma have been injected into the rocks beneath Mammoth Mountain. Carbon dioxide is apparently escaping from this magma and seeping to the surface. Here is a classic standoff between the scientists investigating a restless caldera and the residents of Mammoth Lakes, a town supported by a ski resort and summer recreation. The investigators want to understand

TABLE 5
Estimated gas emissions from world's volcanoes on land

Gas	10^{12} Grams/Year
Water	713
Carbon dioxide	65
Sulfur dioxide	19
Hydrogen chloride	3
Hydrogen fluoride	0.1

Adapted from Richard Stoiber, "Volcanic Gases from Subaerial Volcanoes on Earth," *Global Earth Physics*, American Geophysical Union, Reference Shelf no. 1, 1995.

NNW   SSE

Postcaldera rhyolite
and sedimentary fill   SIERRA
                                                              NEVADA
Precaldera volcanics
                                                    South Moat

                                                              January 1983
Bishop Tuff   earthquake
                                                              swarm

            Sierran basement   Possible
                                                              dike intrusion

Crystallized
magma

                  Possible magma chamber

0   1   2   3   4   5 km
No vertical exaggeration

118   A schematic cross section through Long Valley Caldera near Mammoth Lakes, California. This caldera formed when 600 cubic kilometers of pyroclastic flows called Bishop Tuff were erupted about 700,000 years ago. The most recent eruptions, 500 to 600 years ago, formed a north–south line of rhyolite domes in and north of the caldera. Earthquake swarms and uplift within the caldera from 1980 to 1996 indicate a probable injection of new magma beneath the area. (Adapted from D. P. Hill, R. A. Bailey, and A. S. Ryall, *Journal of Geophysical Research* 90 [1985]: 11111-20 and cover.)

how volcanoes work and reduce volcanic risk; many residents wish they would do it somewhere else.

Recent study and analysis of submarine hot springs that occur along mid-ocean ridges indicate that they also are important sources of volcanic gases, especially carbon dioxide. The mid-ocean ridge lavas contain a low percentage of gas, but their abundance compared with the lavas erupted on land makes them a major supplier of gases to the Earth's surface.

The sum of all these studies allows some reasonable estimates of the volume and composition of the volcanic gases currently being added to the Earth's water and air. Based on the numbers of their atoms, hydrogen is the most important constituent of volcanic gas, followed by oxygen, carbon, sulfur, chlorine, and nitrogen. As the elements combine at the surface of the Earth they become water, carbon dioxide, sulfur dioxide, hydrochloric acid, and nitrogen. The ratios of these volcanic gases are in reasonable accord with the ratios of water, carbon, chlorine, and nitrogen found in the Earth's air, oceans, and surface rocks. However, the ratio of sulfur is not.

Sulfur appears to be a special problem. At the current rate of emission from volcanoes, there should be fifty times more sulfur at the Earth's surface. What has saved us from the sulfuric acid blanket that plagues the planet Venus? The answer is probably iron sulfide—the common mineral pyrite, or "fool's gold"—and also the plate tectonic process. Instead of being released into the ocean or atmosphere, sulfur reacts with iron and forms insoluble iron sulfide, which remains in the oceanic crust and is returned via a subduction zone to the shallow interior of the Earth.

The probability that sulfur is scavenged from the Earth's surface by the seafloor-spreading machine has spawned a new school of thought championed by William Fyfe of the University of Western Ontario in Canada. Fyfe points out that perhaps more water is being consumed in the water-rich rocks entering subduction zones than is being returned to the surface in volcanic gases.

This idea reverses the older concept that the Earth has been releasing gases to the surface throughout geologic time. In this new view, a primitive Earth had a thick atmosphere and universal ocean that covered solid rocks of low water and gas content. The stirring of this Earth by plate tectonics caused the surface water and gases to mix down into the mantle rocks, perhaps scrubbing out the excess sulfur in the process.

Although the compositional ratios appear generally correct for a volcanic recycling of the Earth's air and water, what about the total amounts of these gases? Figure 119 shows the volatile elements and compounds in the Earth's oceans, sedimentary rocks, and atmosphere compared with their volcanic production. From the recycling point of view, this comparison indicates that about 25 percent of the water, chlorine, and nitrogen in the Earth's atmosphere was reworked by subduction and volcanic eruption and about 75 percent of the carbon was recycled.

The data in Figure 119 are still being refined; the total abundances of water, chlorine, carbon, and nitrogen are fairly well established, but sulfur in the top layers of the oceanic crust is not included. The volcanic exhalations are only first approximations, and it is not known

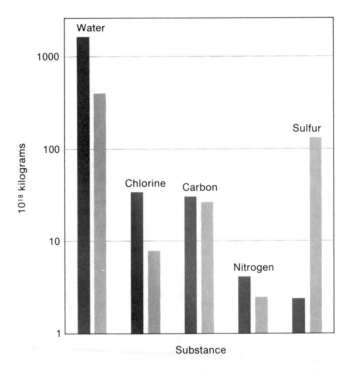

119 A comparison of volatile molecules and elements in the Earth's oceans and atmosphere with their volcanic production. (Note logarithmic scale.) For each substance, the bar on the left represents the total weight of that material in the Earth's air, oceans, and sedimentary rocks. The bar on the right represents the weight of that same substance currently being produced from volcanoes multiplied by the Earth's age. It is apparent that some process such as subduction must remove sulfur from the Earth's surface.

whether the current volcanic activity is representative of volcanism throughout geologic time. Nevertheless, a comparison based on approximate data can be important for qualitative purposes, even though future data may change its quantitative conclusions. In other words, the current data support the concept of recycling, even though the total amount of various volatiles that have been recycled is still uncertain.

If the water that still covers most of our world's surface was not originally part of the Earth, it must have been added early from outside. Comets, made largely of ice, are likely candidates for this extraterrestrial source. Apparently comets still occasionally collide with the Earth, and during the early days of the solar system the number of comets and their collisions is thought to have been much greater.

If comets and not volcanoes provided most of the Earth's early sea and air, volcanologists can no longer claim that these essentials of life were born of fire. Nevertheless, recycling is an important task, especially if the overall process can clear the air of sulfuric acid. Ongoing volcanism and tectonism have helped provide the long-term balance among land, sea, air, and climate that is essential to the evolution of life on Earth.

Sources

Anderson, A. T. "Some Basaltic and Andesitic Gases." *Reviews of Geophysics and Space Physics* 13 (1975).

Gerlach, T. M. "Present Day CO_2 Gas Emissions from Volcanoes." *EOS* (American Geophysical Union) 72 (1991).

Gerlach, T. M. "Volcanic Gases." *Encyclopedia of Geochemistry, Encyclopedia of Earth Science Series*. London: Chapman & Hall, 1997.

Graedel, T. E., and P. J. Crutzen. *Atmospheric Change: An Earth System Perspective*. New York: Freeman, 1995.

Holland, H. D. *The Chemical Evolution of the Atmosphere and Oceans*. Princeton, NJ: Princeton University Press, 1984.

Stoiber, Richard. "Volcanic Gases from Subaerial Volcanoes on Earth." *Global Earth Physics*. American Geophysical Union, Reference Shelf no. 1, 1995.

Symonds, R. B., W. I. Rose, G. J. S. Bluth, and T. M. Gerlach. "Volcanic Gas Studies." Chap. 1 in M. R. Carroll, and J. R. Hollaway, eds., *Volatiles in Magma*. Vol. 30 of *Reviews in Mineralogy*. Mineralogical Society of America, 1994.

14

Volcanic Power

120   The Wairakei geothermal field generates electrical power in New Zealand.

Fires that shook me once, but now to silent
ashes fall'n away
Cold upon the dead volcano sleeps the gleam of
dying day.

—Tennyson (1809–1892)

Geothermal energy was a term known to only a few specialists before the 1974 energy crisis, a crisis that has now almost been forgotten. Although not exactly a household name, it has become a more familiar term as an alternative source of energy.

In discussing geothermal energy and power, it is important to understand the subtle difference between these terms. Heat is a form of energy, and there is an immense amount of it inside the Earth. For each kilometer of depth the temperature increases about 20° to 60°C, depending on the region (Figure 121). The heat energy contained in just the upper 10 kilometers of the United States is estimated to be 3.3×10^{25} joules, equal to the heat released by burning 1 thousand trillion tons of coal. If it were accessible, it would supply our energy needs for the next 100,000 years.

But potential energy becomes practical only when it can be consumed at some useful rate. *Power* is a measure of the rate at which energy is made available or is consumed. Energy is measured in joules, and power is measured in joules per second, or watts. There is plenty of diffuse energy in the Earth; turning it into useful power is the critical problem.

The difference between energy and power becomes clear in the following examples: The amount of solar energy is enormous, but the solar power reaching the Earth is modest—averaging about 40 watts for each square meter of surface. A lightning bolt contains only a modest amount of energy, equal to that of about two barrels of crude oil, but it expends this energy in a fraction of a second, unleashing enormous power.

Although volcanic explosions are often compared with nuclear bomb blasts, the analogy is misleading. The heat and mechanical energy in the

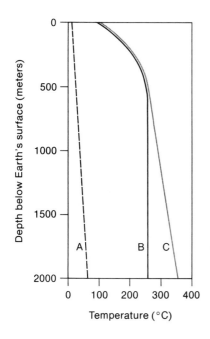

121   Temperatures beneath the Earth's surface in regions of hot springs and geysers are distinctly different from those in more normal regions (Curve A). Curve B is typical of temperatures in a geyser basin or a geothermal steam field. Curve C is the temperature of boiling water under hydrostatic pressures at various depths. (Adapted from D. E. White, U.S. Geological Survey.)

1883 eruption of Krakatau was on the order of 5000 megatons of TNT, but this energy was released in a series of explosions lasting almost a day. A nuclear bomb, however, goes off in a single, almost-instantaneous flash, unleashing power in amounts unmatched by geological phenomena.

Geothermal energy—like solar energy—is enormous, but its natural rate of release is small, averaging about one-sixteenth watt from beneath each square meter of the Earth's surface. Even if geothermal energy could be converted to electricity with an efficiency of 20 percent, it would require all the heat flow from an area as large as a football field to power one 60-watt lightbulb. Only in a few volcanic regions like Yellowstone, New Zealand, and Iceland, where geysers and hot springs are abundant, does nature sustain a significant output of geothermal power (Figure 120 and Color Plates 13 and 24).

For geothermal power to be practical, some special situation must exist that can concentrate the Earth's heat energy into a small area. Natural or artificial underground reservoirs of steam or hot water that can be funneled into a drill hole provide this special situation (Figure 122).

In nonvolcanic areas, holes drilled to depths of 4 or 5 kilometers may reach temperatures near 100°C. However, in young volcanic regions, where molten rock has brought heat up from deeper levels, it is possible to drill into rocks and steam reservoirs at depths of 1 to 3 kilometers that are heated to 100° to 350°C or hotter. To develop the true potential of geothermal power, it is necessary to locate these shallow

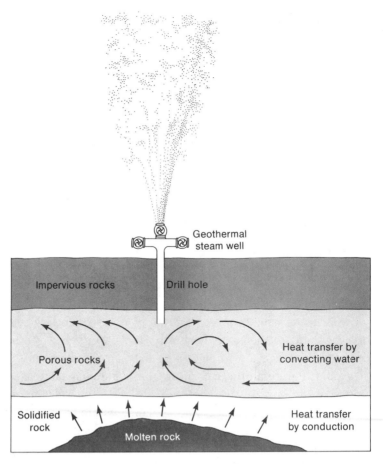

122   A schematic diagram of a geothermal power reservoir.

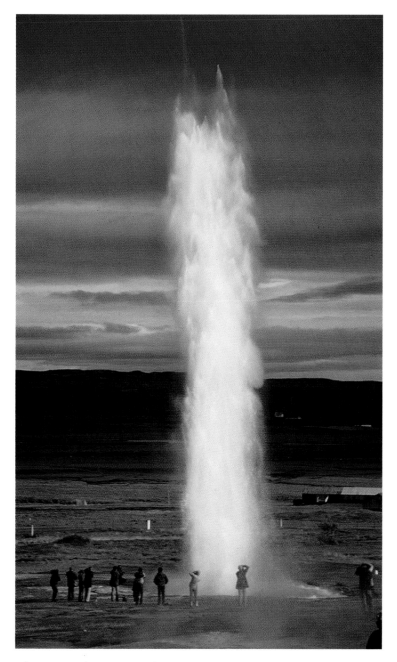

Plate 13 The eruption of a geyser in central Iceland. *Geysir,* an Icelandic word meaning "to gush," is the original word for these spectacular spouting hot springs. Other noteworthy examples can be found in Yellowstone National Park and near Rotorua in New Zealand.

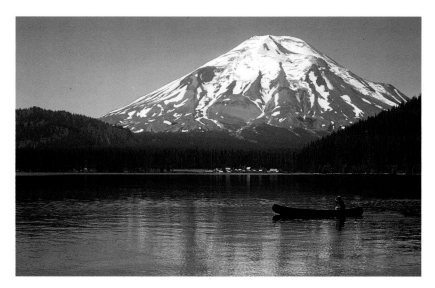

Plate 14   Mount St. Helens in 1969, viewed from Spirit Lake. Before the massive eruption in 1980, the summit elevation of this beautiful stratovolcano was 2950 meters.

Plate 15   Mount St. Helens and Spirit Lake in 1982. The rim of the 2-kilometer-wide, horseshoe-shaped crater reaches an elevation of 2549 meters, and the crater bottom is more than 1000 meters below the pre-1980 summit elevation. The new lava dome can be seen inside the crater. (Photograph by Lyn Topinka, U.S. Geological Survey)

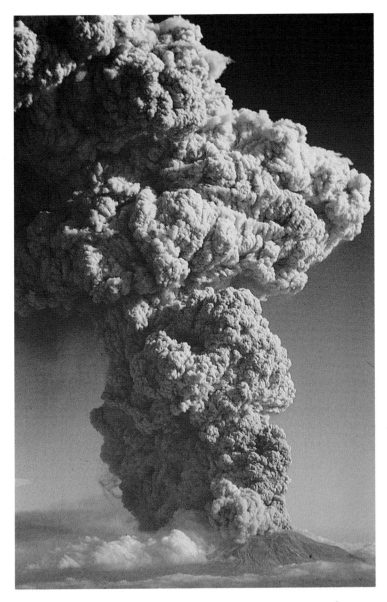

Plate 16 The explosive eruption of Mount St. Helens on July 22, 1980. Although this explosion generated an ash cloud that reached a height of 14 kilometers, its energy was about one hundred times less than that of the May 18, 1980, eruption. (Photograph by Katia Krafft)

Plate 17 The eruption cloud from Mount Pinatubo in the
Philippines on June 12, 1991. The climactic eruption occurred on
June 15, 1991. (Photograph by David Harlow, U.S. Geological Survey)

Plate 18 The main lava flow from Mauna Loa's 1984 eruption
poured from a vent at 2900 meters and traveled 25 kilometers down the
mountain in five days to an elevation of 900 meters, only 6 kilometers
from the city of Hilo. (Photograph by David Little)

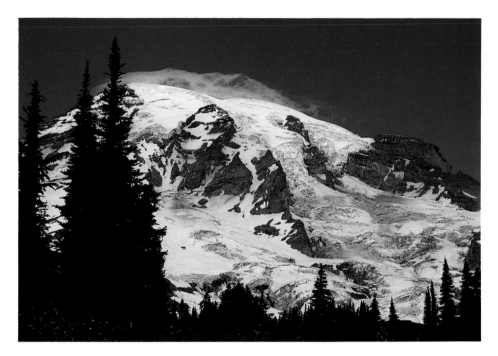

Plate 19 Mount Rainier, a 4393-meter-high, dormant stratovolcano in Washington State. The twenty-five glaciers on Rainier contain 4 cubic kilometers of ice that may cause a debris avalanche and mudflow if an eruption or large earthquake should take place.

Plate 20 A fumarole (volcanic gas vent) in Kilauea Caldera deposits sulfur as the gases cool and mix with air. (Photograph by R. L. Christiansen, U.S. Geological Survey)

Plate 21 Halemaumau Crater in the Kilauea Caldera begins to fill with a lava lake during an eruption in 1961. (Photograph by L. R. McBride, U.S. National Park Service)

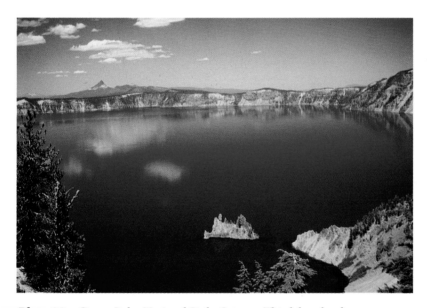

Plate 22 Crater Lake National Park, Oregon. This lake, the deepest in the United States, fills a 10-kilometer-wide caldera that was formed by the great eruption and collapse of Mount Mazama Volcano about 7700 years ago.

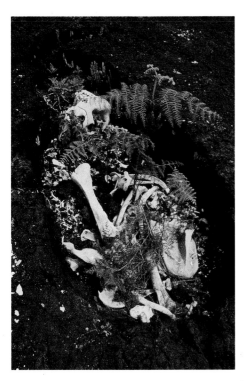

Plate 23 In 1977, a fast-moving lava flow from an eruption of Nyiragongo Volcano in Africa covered part of the mountain's flank, trapping and killing a small herd of elephants. This mold of a dead elephant—note its trunk at the upper left—was preserved in the hardened lava, which forms a sarcophagus for its bones. (Photograph by Katia Krafft)

Plate 24 At Waimangu Hot Springs near Rotorua, New Zealand, terraces of silica are deposited as the water cools and precipitates some of its dissolved compounds. Most of the colors are formed by algae that thrive in the warm, nutrient-rich water.

Plate 25 Submarine hot springs have been discovered at many
locations along the volcanic mid-ocean ridges. These "black smokers" at
13° north latitude on the East Pacific Rise are pouring 350° C mineral-
laden water into the cold, deep, bottom water 2500 meters below sea
level. Iron sulfide and other metallic sulfide particles precipitating
in the suddenly chilled hot-spring water form the black "smoke."
(Photograph by J. L. Cheminée, Observatoires Volcanologiques,
Institut de Physique du Globe de Paris, doc. Ifremer; taken from
the French submersible CYANA in 1982)

bodies of hot rocks and fluids and to tap their pent-up energy through a system of drill holes. Exploring for geothermal power is similar to exploring for oil, except that the geologic setting is volcanic rather than sedimentary. Most of the existing high-temperature geothermal power fields have been found in active or young volcanic regions.

The form of heat transfer in a given region is of key importance. Heat is transferred by radiation, conduction, and convection. *Radiation* transfers a significant amount of heat through transparent media like air and space, but it is much less effective in solids. *Conduction* is such a slow process in good thermal insulators like rocks that it transfers only small amounts of heat from inside the Earth. The very slow release of the Earth's interior heat through its surface (averaging 0.06 watt per square meter) is controlled by this slow conduction process. *Convection* is the only process fast enough to transfer the Earth's heat at rates sufficient to produce significant power. Convection involves the actual physical movement of molten rock or hot fluids, generally upward, because these hot rocks and fluids are less dense than the cooler rocks and fluids surrounding them (Figure 123).

Geothermal power fields are found where magma has moved upward from depths of 50 to 100 kilometers and brought the high temperatures (900° to 1200°C) of these depths to near the surface (Figure 124). Groundwater heated by these volcanic intrusions can form another convection circuit, bringing hot springs and geysers to the surface or remaining sealed below the surface awaiting the wildcatter's drill.

Geothermal prospects related to volcanic activity are of three types: hydrothermal reservoirs, hot dry rocks, and magma reservoirs. Of the three, hydrothermal reservoirs are the easiest to develop because their hot water and steam are waiting to be tapped.

Larderello, Italy, is an area of natural hot springs. The boric acid concentrated in these thermal waters has been extracted for more than 200 years. Wells drilled to increase chemical production encountered steam and were first used to produce electricity in 1904. Some 300 wells have been drilled to depths of 300 to 700 meters over the years and have outlined a reservoir producing steam at 235°C and 30 bars of pressure (30.6 kilograms per square centimeter). The reservoir is in porous

123   A sketch of a woodstove shows heat transfer by convection of the heated air and wood smoke in the stove; by conduction through the firebrick and iron walls of the stove; and by radiation from the surface of the stove through the open space of the room.

limestone, but intrusions of young volcanic rocks in the area are considered to be the heat source. The field has produced some 400 megawatts of power for the last fifty years and is still going strong (Figure 125).

In 1925, Icelanders began to drill for hot water and steam. By now they have developed more than 250 fields whose power is used mainly for space heating. Almost all of Reykjavík, the capital city of nearly 150,000 people, gets its heat and hot water from more than 100 wells, 300 to 2200 meters deep, that tap water at about 100°C in porous basalts. Although once-flowing wells now need to be pumped, the temperatures have shown little decrease. For the people in oil-barren, arc-tic Iceland, geothermal power is the very lifeblood of their warmth and comfort (Figure 126).

In 1921, The Geysers in California—a region of young volcanic rocks in the Coast Range north of San Francisco—was drilled for steam. Electricity was first generated in 1959, and the power field has been greatly expanded since that time. More than 600 wells, 500 to 3200 meters deep, have produced steam at about 240°C and 30 bars of pressure from a reservoir of fractured shaley sandstone. Production was

Key

Regions favorable for geothermal prospecting

● Location of geothermal power plants

124 A map of the world's geothermal regions. Notice the close relationship to zones of active volcanoes.

expanded during the 1980s to nearly 2000 megawatts, enough electricity for a city of 2 million people (Figure 127). In recent years, however, production has declined, down to 1200 megawatts by 1993. Once considered a nearly inexhaustible source of power, the rapid extraction of huge amounts of steam has lowered the field's pressure. Even so, the 1000 megawatts of electricity currently generated at The Geysers makes it the world's largest geothermal power field, and new ideas for water injection and limiting steam-production rates will probably extend its useful life for several decades.

125   Larderello, Italy, is the site of the world's first geothermal steam field. Started in 1904, it now produces about 300 to 400 megawatts of electricity. The huge chimneys are cooling towers similar to those used at nuclear power plants for recycling the cooling water. (Photograph by Patrick Muffler, U.S. Geological Survey.)

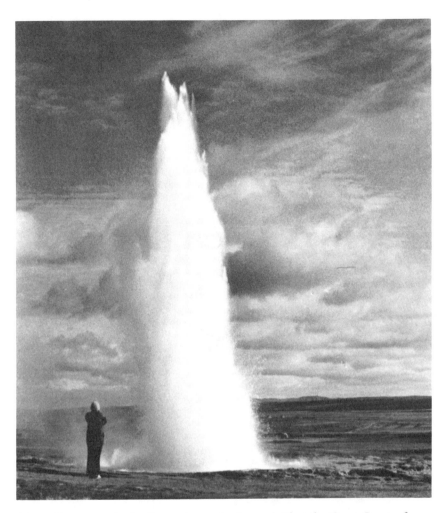

126 *Geysir* is an Icelandic word meaning "to gush," but the Great Geysir of Haukadalur, the world's original geyser, seldom performs. Its small neighbor, Strokkur Geysir, seen in this photograph, erupts every few minutes.

On the island of Hawaii, a geothermal exploration well encountered some of the hottest underground fluids found so far. Drilled into lava flows on the east rift of Kilauea Volcano to a depth of 1969 meters, the hole had a bottom temperature of 350°C. This discovery well generated

127   The Geysers, California, once a hot-spring health spa, is now the world's largest geothermal steam field. Producing nearly 1200 megawatts, the field generates enough electricity to supply a city of 1 million people. (Photograph by Pacific Gas and Electric Company.)

about 2 to 3 megawatts of electricity during the 1980s (Figure 128). Additional wells have been drilled nearby, and a 25-megawatt generating plant is currently producing a significant part of the electricity consumed on the Big Island.

The prospects for generating geothermal power from hot dry rocks are more speculative. These potential fields involve areas of higher-than-normal thermal gradients on the order of 60°C per kilometer of depth in areas of impermeable rocks. Holes drilled to 3- or 4-kilometer depths can reach down into these hot dry rocks. Artificial fractures are established between two closely spaced holes, and cold surface water is pumped down an injection well. Heated by the hot rock along the fractures, this water rises as hot water or steam in a return well (Figure 129).

Los Alamos National Laboratory in New Mexico performed a hot-dry-rock experiment on the flank of the Valles Caldera during the 1980s. Cold water pumped down a 3-kilometer-deep well returned at 135°C up another well close by. However, the cost of the experiment far exceeded the value of the power produced. Its proponents declared the experiment a scientific success, and its opponents declared it a commercial failure. As long as relatively inexpensive oil and natural gas are available, the concept of power from hot dry rock will remain an idea whose time has not yet come. Nevertheless, its potential is enormous.

128   Hawaii's first geothermal steam well is located on the east rift of Kilauea Volcano. A 3-megawatt electrical generator was installed on this well. (Photograph by Larry Kadooka, *Hawaii Tribune Herald.*)

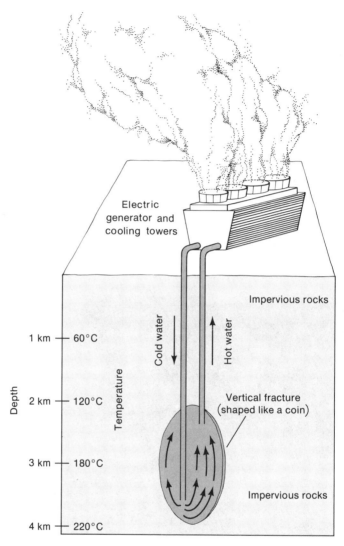

129 A schematic diagram of the hot-dry-rock geothermal power concept. Two holes are drilled into impervious rocks in regions where the temperature increases rapidly with depth. The holes are connected by hydraulic fracturing, a process of fracturing rock by pumping high fluid pressure down a well. Surface water is injected into one hole; it is heated from contact with hot rocks at depth and rises in the second hole as hot water or steam. Cooling of the rocks at depth causes further fracturing, allowing the circulating water to reach additional hot surfaces. Preliminary tests near Los Alamos, New Mexico, with wells drilled to a depth of 3 kilometers in granite at 200°C, have been encouraging.

Much of the young mountain terrain in the western United States, as well as that in Hawaii and Alaska, is of volcanic origin and forms a tempting but well-buried treasure of geothermal energy.

Even molten rock itself is a potential source of power. Magma contains about 1000 joules of heat energy per gram. The energy in 1 cubic kilometer of magma is enough to light San Francisco for 200 years. But tapping it requires truly advanced technology: although scientifically it is feasible to extract power directly from shallow bodies of magma beneath volcanic centers, the engineering technology is still years away.

Initial experiments that involved drilling into buried masses of molten rock were conducted on a lava lake in Kilauea Iki Crater in Hawaii. This great lake of lava formed during an eruption of Kilauea Volcano in 1959, when lava flows ponded in an ancient crater to a depth of 100 meters. A solid crust soon formed on the lake, but lava is such a good insulator that holes drilled in 1979 deeper than 50 meters were still reaching partially molten rock. Conventional rock-drilling techniques using abundant water to keep the drill bit cool were successful in reaching the molten rock, although efforts to place devices down the holes into the molten rock to determine the feasibility of extracting energy were not successful.

Even though geothermal power is still a young and largely unproven industry, its potential makes it worth serious effort and investment. In an assessment of geothermal power in the United States, the U.S. Geological Survey made the following estimates: (1) Identified resources of geothermal reservoirs (excluding National Parks) exceeding 150°C are sufficient to generate 23,000 megawatts of electricity for thirty years, and (2) geothermal reservoirs of less than 150°C are sufficient to generate 280,000 megawatts of thermal power for hot water and heating for thirty years. Undiscovered resources could increase these estimates severalfold. Most geothermal fields have been discovered by drilling where there are surface thermal areas such as hot springs, but the number of hidden fields is difficult to estimate. (Table 6 lists the world's geothermal power production as of 1994.)

Another asset of geothermal power is that it is relatively free of pollution compared with the burning of fossil fuels. Geothermal steam and

TABLE 6
Geothermal power plant capacity in 1995

Country	Megawatts of Electricity
Argentina	0.67
Australia	0.17
China	28.78
Costa Rica	55
El Salvador	105
France (Guadeloupe)	4.2
Iceland	49.4
Indonesia	309.75
Italy	631.7
Japan	413.71
Kenya	45
Mexico	753
New Zealand	286
Nicaragua	35
Philippines	1227
Portugal (Azores)	5
Russia	11
Thailand	0.3
Turkey	20.6
United States	2816.7
Total	6798

Adapted from Gerald Huttrer, "The Status of World Geothermal Power Production 1990–1994." Proceedings of World Geothermal Congress, Florence, Italy, 1995.

hot water do not produce soot and usually contain only small amounts of carbon dioxide, one of the "greenhouse" gases that may cause global warming. As world population grows and world energy resources decline, the geothermal power industry will certainly become an important component in sustaining civilization as we know it.

Volcanoes, directly and indirectly, are powerhouses of enormous potential. The most scenic, such as Yellowstone, Hawaii Volcanoes, and other volcanic national parks, should never be developed (Figure 130). Other volcanoes may help provide light and warmth in centuries to come, but they will require a major research and engineering effort involving much time, labor, and money to find out. As with all human adventures, high payoff involves high risk.

130   Old Faithful geyser in Yellowstone National Park has erupted every 40 to 80 minutes for the past 100 years. Spraying 50,000 kilograms of boiling water 35 to 50 meters high, each eruption releases about 1.9×10^8 joules of energy during its few minutes duration (about 1 megawatt of power). Thick deposits of silica precipitated from the cooling waters indicate that the Yellowstone geyser basins have been active for thousands of years.

Sources

Duffield, Wendell, J. H. Sass, and M. L. Sorey. *Tapping the Earth's Natural Heat*. U.S. Geological Survey Circular no. 1125, 1994.

Huttrer, Gerald. "The Status of World Geothermal Power Production 1990–1994." Proceedings of World Geothermal Congress, Florence, 1995.

Muffler, L. J. P., ed. *Assessment of Geothermal Resources of the United States—1978*. U.S. Geological Survey Circular no. 790, 1979.

Rybach, L., and L. J. P. Muffler. *Geothermal Systems: Principles and Case Histories*. New York: Wiley, 1981.

Wohletz, K., and G. Heiken. *Volcanology and Geothermal Energy*. Berkeley and Los Angeles: University of California Press, 1992.

15

Volcanic Treasures

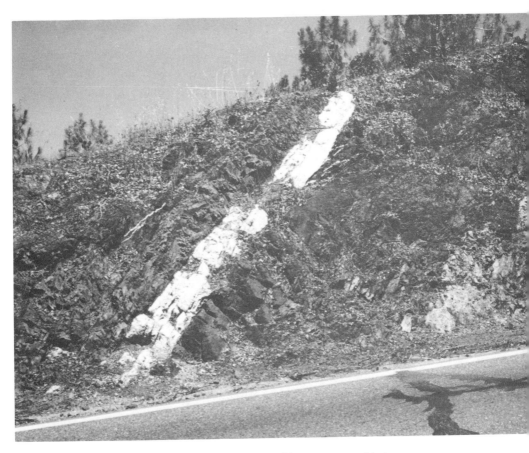

131 Quartz vein in a road cut, in Mother Lode gold country near Mariposa, California.

Volcanoes are nature's forges and stills where the elements of the Earth, both rare and common, are moved and sorted. Some elements are diluted and some pass through unchanged, but many are transported and concentrated into those precious lodes that people seek for fortune or industry.

Economically important elements concentrated directly by volcanic action, especially by the intrusion of magma bodies into the Earth's crust, include fluorine, sulfur, zinc, copper, lead, arsenic, tin, molybdenum, uranium, tungsten, silver, mercury, and gold. These concentrations are mainly formed as hydrothermal vein deposits, which are the mineral fillings precipitated from hot waters percolating along underground fractures. Veins generally consist of one or more common minerals, like quartz or calcite (calcium carbonate), in which the more precious minerals, like gold or galena (lead sulfide), are scattered as small specks or crystals.

Theoretically the process is simple, but in operation it can be enormously complex. Basically, the magmatic roots of volcanoes supply the heat source and perhaps some of the ingredients for a giant still. As the magma cools and the common silicate minerals crystallize to form gabbro or granite, the water and other gases—as well as the rarer elements that do not fit into the rock-forming silicate minerals—become concentrated in the residual silicate melt. As cooling reduces the volume of the rocks they crack, allowing the hot residual magmatic fluids, rich in water and precious elements, to escape from their underground forge. During the ascent of these hydrothermal solutions toward the surface, cooling and decreasing pressure cause various minerals, both common and rare, to precipitate and form veins. Certain minerals precipitate over a large range of pressures and temperatures and are common throughout the vein; others such as gold and silver may precipitate over a very narrow range of pressures and temperatures to form localized bonanzas within the vein (Figure 132).

Veins are usually steeply inclined planes dipping into the Earth, a few centimeters to many meters thick and often many hundreds of meters or several kilometers in length (Figure 131). The gold-bearing veins of the Mother Lode system in the Sierra foothills of California trend north–south and dip steeply down toward the granite roots of the range. Although each vein is generally no more than a meter or so thick and a few kilometers long, the vein system crops out for more than 300 kilometers in the western Sierra Nevada. Hot waters escaping through steep cracks from the volcanic roots of the range deposited mainly hard, milky white quartz in the veins, with few (but enough) specks of gold and pyrite (iron sulfide, or fool's gold) to start the rush to California.

Hydrothermal fluids rise through fractures and fissures and migrate along bedding planes

Some reach surface as geysers or hot springs

Intruded, deformed rocks

Disseminated deposits form by fluids invading small cracks or permeable zones adjacent to veins

Solidified magma

132 Ore deposits are often found close to an intrusion of solidified magma. Many such deposits probably precipitate out of the hot waters associated with the cooling magma. (From F. Press and R. Siever, eds., *Earth*, 2nd ed., p. 577. W. H. Freeman and Company, Copyright © 1978.)

The type of rock surrounding the cooling magma reservoir is important to the formation of ore deposits. In Hawaii (in fact, in most oceanic volcanic islands) ore deposits are nearly nonexistent. Part of the reason for this may be that most oceanic islands are slowly sinking under their own enormous load, thus hiding any ore deposits that have formed. But even in the unusual case where uplift and erosion have exposed the roots of volcanic islands, hydrothermal ore deposits are rare or absent. Volcanic rocks richer in silica than basalt—for example, dacite and rhyolite—are more often associated with hydrothermal ore deposits. Many geologists believe the reason is that the rocks surrounding the magmatic roots of volcanoes are the true source of the valuable elements found in hydrothermal veins. According to this view, the volcanic rocks act as a heat source that pumps existing groundwater into a giant circulating system. Cold waters, being heavier, move down and into the cooling volcanic rocks, carrying trace quantities of valuable elements leached from the surrounding rocks. Heated by the cooling magma, they become less dense and rise into the fractured rocks above. Cooling and losing pressure again, the waters precipitate their quartz and precious ores into the veins forming above the volcanic hearth (Figure 133).

Because mercury, copper, sulfur, and fluorine have been measured in volcanic gases in Hawaii, there is no question that at least some of these elements originate directly in volcanoes. Gold, silver, and other valuable elements found in hydrothermal ore deposits may have to have been concentrated beforehand in some earlier generation of continental crustal rocks before they can be reconcentrated by volcanic distillation.

Two interesting discoveries about the relation of gold to subduction volcanoes have been reported. Geologists studying the fumarole and hot-spring chemistry of Galeras Volcano in Colombia have found that one-half kilogram of gold is escaping to the atmosphere every day from the gas vents, and that more than 20 kilograms of gold per year are being deposited inside the volcano. Of course these emanations and deposits are very low grade, but if all the gold could somehow be collected, it would amount to more than $2 million worth per year.

The other startling find is a major gold discovery that apparently is still being deposited by a shallow, active geothermal system on Lihir, a volcanic island in Papua New Guinea. The deposit lies in a caldera on this young volcano. Speaking in geologic time, of course, it has been said that today's ore deposits were often yesterday's geothermal systems,

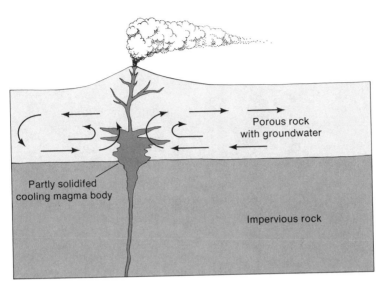

133 A schematic diagram of groundwater being circulated by the cooling of a magma body. The water seeps into the cooling magma body, is heated, and convects upward. Reaching cooler rocks above the magma body, the circulating groundwater also cools and precipitates vein minerals. The source of the minerals may be either the magma or the porous rock through which the groundwater circulates. In this model, the cooling magma body acts as a circulating pump and concentrator rather than as a direct source of ore deposits.

but here is a place where both are present today. It is ironic that in order to mine the gold at Lihir, the geothermal system will somehow have to be cooled first.

Each mineral has its characteristic place in the hydrothermal system. Tungsten minerals precipitate at very high temperatures and often are found at the very point of contact between a chilled magma body and the rocks it has invaded, especially limestone. Scheelite, a calcium tungstate mineral, is brightly fluorescent under ultraviolet light. Tungsten prospectors seek out contacts between granite and limestone during the day and then prospect these outcrops at night with "blacklights" (ultraviolet lamps) to reveal the areas of high tungsten concentration.

Mercury is at the other extreme. Both as an element and in compounds it is volatile; that is, it forms a gas at low temperatures and pressures. For this reason, much of the Earth's mercury is lost to the surface from volcanic steam vents and hot springs. Near many active volcanoes the concentration of mercury in the air exceeds the health standards established by environmental protection agencies, but it is difficult to get the volcanoes to stop smoking.

Hydrothermal ore deposits in continental locations have been studied by geologists for more than 100 years, and were mined for thousands of years before that, but only recently has the importance of submarine ore bodies deposited at mid-ocean rifts been recognized.

In the 1960s scientists probing the floor of the Red Sea discovered some areas where hot but dense, highly saline bottom water overlies muds that are rich in iron, zinc, and copper minerals. Since the Red Sea is a volcanic rift zone, oceanographers suspected that the metal-rich sediments might have been precipitated from submarine hot springs.

Proof of this idea came in the late 1970s, when scientists in submersibles peered out on the strange landscapes of volcanic rift zones in the eastern Pacific Ocean. Twenty-five hundred meters below the surface, they were startled to see the "black smokers" that we discussed in Chapter 3—chimneys of iron, zinc, and copper sulfides a few meters high, pouring out dark particles of iron sulfides. The water spouting from these chimneys was hot—350°C—and as it was quickly chilled by the surrounding seawater, the dissolved metal sulfides precipitated into fine particles (the black smoke) or into the chimneys. Some of the metal sulfide deposits contain small but significant amounts of silver and gold. In addition to building the chimneys, the ore deposits form mats and crusts as much as a few meters thick that are spread over several hundred square meters around the vents.

The general model for the operation of these submarine hot springs and their attendant ore deposits is shown in Figure 134. Heat from the magma chamber beneath the rift zone drives a convective circulation of seawater within the porous basalts of the seafloor. Water sucked in on the rift flanks is heated and dissolves calcium, silica, and metallic sulfides while depositing magnesium and sulfate. The heated discharge waters—containing volcanic fluids rich in carbon dioxide and hydrogen sulfide—are cooled as they emerge on the seafloor and deposit their metal sulfides.

The large extent of submarine ore deposits is just now being recognized. They apparently occur in many places along mid-ocean ridges, and as the seafloor spreads away from the ridge, the areas of ore bodies

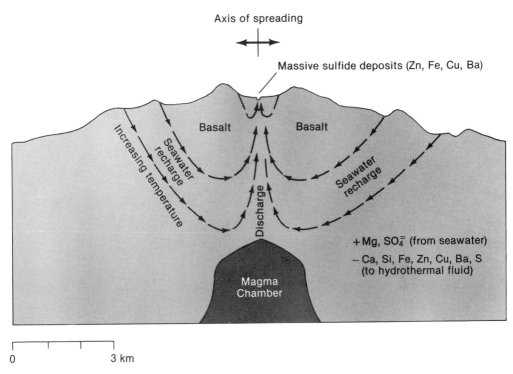

Axis of spreading

Massive sulfide deposits (Zn, Fe, Cu, Ba)

Basalt   Basalt

Increasing temperature

Seawater recharge

Seawater recharge

Discharge

+ Mg, $SO_4^=$ (from seawater)

− Ca, Si, Fe, Zn, Cu, Ba, S (to hydrothermal fluid)

Magma Chamber

0   3 km

134 A schematic cross section of fluid circulation and hydrothermal deposits at a submarine rift zone. The seafloor topography is exaggerated, and the depth to the magma chamber is uncertain. (R. A. Koski, U.S. Geological Survey.)

increase. Although they are currently so deep that they are out of reach of commercial mining, a few ancient deposits have been shoved up on land where oceanic plates have converged on continents. The metal sulfide mines in Cyprus and Japan, long puzzling to geologists, appear to have originated at submarine hot springs. Most ancient submarine ore deposits, however, probably have been consumed in subduction zones as the ocean plates bearing them pushed beneath a continent. Perhaps they were and still are being recycled by subduction volcanoes into the more conventional hydrothermal ore deposits found in continental rocks.

Diamonds, perhaps the greatest mineral treasure of all, are also closely related to volcanic processes. Diamonds and the graphite cores of common pencils are the same element—carbon—but they are different minerals. In graphite, the carbon atoms are arranged in layers like the silica atoms in sheets of mica, and the loose bonding between layers permits the sheets to break and slide past one another. Graphite is therefore soft and greasy feeling. In diamonds, the atoms of carbon are compressed into a tight network that interlocks in all directions, forming the hardest substance known on the Earth (Figure 135). To achieve this close packing of carbon atoms, extremely high pressures are needed—pressures that occur naturally only at depths of nearly 200 kilometers inside the Earth. Once formed, diamonds are stable at low pressures and temperatures, but at the Earth's surface they can burn at high temperatures.

Diamond

Flaky cleavage

Graphite

135 The crystal form and internal structure of diamond and graphite. The three-dimensional network of atomic bonds in diamond makes it extremely hard and durable. In contrast, the two-dimensional network of major bonds in graphite makes it flaky and soft. (From J. Gilluly, A. C. Waters, and A. O. Woodford, *Principles of Geology*, 4th ed., p. 30. W. H. Freeman and Company, Copyright © 1975.)

The right conditions to form diamonds apparently exist beneath the continents at a depth of nearly 200 kilometers and at a temperature near that of molten rock (Figure 110). In fact, diamonds might not be all that rare if we could mine them at their deep source. Their occurrence at the Earth's surface results from a rare type of volcanic eruption that transports them rapidly from great depths into shallow vents called *Kimberlite pipes* (Figure 136). The pressure and temperature are reduced so quickly that the diamonds do not revert to a more normal surface form of carbon such as graphite.

No one knows whether the type of volcano associated with diamond pipes exists today. The peculiar Kimberlite volcanic rock and the presence of other minerals indicating that the volcanic fluids were extremely high in carbon dioxide suggest an unusual kind of explosive volcanism that may occur only during certain episodes of geologic history. A few active volcanoes in East Africa, especially those whose rocks are unusually high in carbon dioxide (like Ol Doinyo Lengai), may be the closest active relatives of a diamond-pipe volcano (Figure 137).

On an expedition to the summit of Ol Doinyo Lengai, scientists found small, active lava lakes in the crater. The temperatures of these molten lavas ranged from 500° to 550°C and appeared to have very low viscosity. The exceptionally low temperatures—about half that of basaltic lava—apparently result from the high sodium carbonate content of the Ol Doinyo Lengai lavas. The soda acts as a flux, reducing the melting temperature in the same way it does in making commercial bottle glass.

Even if they are not transporting diamonds, most of the world's active volcanoes are probably forming some type of hydrothermal deposits beneath their surface today. This is also true of geothermal reservoirs in which the circulating underground hot water is selectively dissolving, transporting, and depositing the more soluble minerals within the porous rocks. On a geologic time scale, today's volcanoes and geothermal systems are tomorrow's ore deposits.

The fact that a natural geothermal area like Yellowstone has a life span of tens of thousands to hundreds of thousands of years indicates that volcanic ore deposits form quite slowly—drop by drop, atom by atom. People and their machines have voracious appetites for minerals, consuming them at rates far in excess of nature's patient creation. The message here is clear: we must recycle whatever mineral wealth we can or wait eons for new supplies.

Lower Jurassic

Triassic

Pipe prior to erosion

Permian

Present land surface

Mined-out zone

Sill

Shales

Lavas

Quartzites

Conglomerate

Precambrian

Gneisses and schists

Granite

Limit of mining

136 A. The famous Kimberley Diamond Pipe in South Africa. This volcanic vent was mined more than 1000 meters deep before 1908. (Photograph by DeBeers Consolidated Mines Limited) B. The cross section shows a reconstruction of the pipe before it eroded, the present ground surface, and the mined-out zone. (Diagram after Arthur Holmes.)

137 Ol Doinyo Lengai, "Mountain of God," in Tanzania, East Africa. The high content of sodium carbonate in lavas from Ol Doinyo Lengai is unique, suggesting that it may be related to the type of volcanism that forms diamond pipes. (Photograph by Richard Stoiber.)

Sources

Boyd, F. R., and Henry 0. A. Meyer, eds. *Kimberlites, Diatremes, and Diamonds: Their Geology, Petrology, and Geochemistry*. Washington, DC: American Geophysical Union, 1979.

Goff, Fraser, J. A. Stimac, A. C. L. Larocque, J. B. Hulen, G. M. McMurty, A. L. Adams, A. Roldan-M, P. E. Trujillo Jr., D. Counce, S. J. Chipera, D. Mann, and M. Heizer. "Gold Degassing and Deposition at Galeras Volcano, Colombia." *GSA Today*, October 1994.

Guilbert, J. M., and C. F. Park. *The Geology of Ore Deposits*. New York: Freeman, 1986.

Herzig, Peter, M. Hannington, B. McInnes, P. Stoffers, H. Villinger, R. Seifert, R. Binns, and T. Liebe. "Submarine Volcanism and Hydrothermal Venting Studied in Papua New Guinea." *EOS* (American Geophysical Union), October 11, 1994.

Humphris, Susan, R. A. Zierenberg, L. S. Mullineaux, and R. E.Thomson, eds. *Seafloor Hydrothermal Systems*. American Geophysical Union Geophysical Monograph no. 91, 1995.

Koski, Randolph A., et al. "Hydrothermal Processes and Massive Sulfide Deposits on the Juan de Fuca Ridge and Other Northeast Pacific Spreading Axes." *Circum Pacific Council for Energy and Mineral Resources Earth Science Series* 6 (1987): 621–638.

Rona, Peter. "Mineral Deposits from Sea-Floor Hot Springs." *Scientific American*, January 1986.

16

Volcanoes and Climate

138　Vesuvius, 1944. (Photograph by the U.S. Navy, courtesy of the National Archives.)

*In one period we believe ourselves governed
by immutable laws; in the next by chance.*

—Loren Eiseley (1907–1977)

B ad weather has been blamed on almost everything—from atomic bombs, sun spots, and the Industrial Revolution to volcanoes, black magic, the Republicans, and the Democrats. And even though most bad weather probably results from the natural variability of atmospheric processes, at least two of these culprits—the Industrial Revolution and volcanoes—have put large amounts of debris and gas into the atmosphere and thus appear more suspect than some of the others (Figure 138).

Benjamin Franklin was the first to suggest that volcanoes modify climate and weather. When Laki Volcano in Iceland erupted in 1783 with the largest effusion of lava in history, an enormous amount of gas was released, enveloping Iceland and much of northern Europe for months in a blue haze or "dry fog" (Figure 139). The gas must have contained a significant amount of fluorine because livestock grazing on contaminated grass in Iceland died of fluoridosis. The widespread death of livestock—11,000 cattle, 28,000 horses, and 190,000 sheep—resulted in a severe famine in which 10,000 Icelanders, one-fifth of the population, perished. By the time the "dry fog" reached Europe it was more annoying than poisonous, but it was visible on many days during the summer and fall of 1783 and was apparently observed by Franklin during his stay in France. Since the winter of 1783–1784 was abnormally severe, especially in Europe, Franklin suggested that the fine ash and gases from the Laki eruption may have filtered out enough of the sun's rays to cause the cold weather.

In 1815 Tambora Volcano shook the island of Sumbawa in Indonesia with a gigantic explosive eruption. Recent expeditions to Tambora estimated that the amount of magma expelled in high ash clouds and pyroclastic flows during that great eruption amounted to about 40 cubic kilometers. Although there is little data on the actual atmospheric effects from Tambora's huge explosion and caldera collapse, 1816 was

139 Cinder cones along the Laki fissure in Iceland. This fissure erupted in 1783 with the greatest lava flood in recorded history. The gases released during this enormous eruption caused a blue haze or "dry fog" that reached Europe.

remembered in Europe and America as the "year without a summer"— a year when snow fell in New England in June.

The idea that a volcanic eruption could affect the world's climate surfaced again after the 1883 eruption of Krakatau (Chapter 4). This time the visible atmospheric effects were unquestionably worldwide; they began within two weeks of the great explosive eruption and lasted for months. Strange colors and halos around the sun and moon were noted, and there were vivid sunrises and sunsets for months on end (Figure 140). A Ceylon (now Sri Lanka) newspaper for September 17, 1883, gave this account:

A

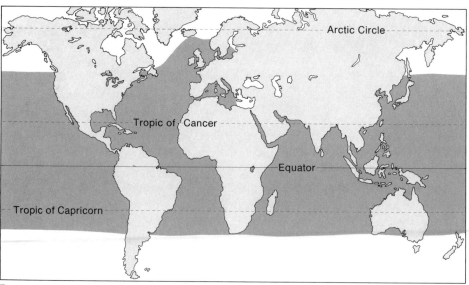

B

140 *(Facing page)* A. The approximate distribution of sky phenomena between August 26 and September 7, 1883. B. The approximate limits of the main sky phenomena at the end of November 1883. High-speed stratospheric winds averaging nearly 120 kilometers per hour carried the fine volcanic dust of the 1883 Krakatau eruption westward around the globe. By the end of November, the stratospheric haze covered over 70 percent of the Earth's surface, causing spectacular sunsets and strange optical phenomena. (Adapted from C. J. Symons, ed., *The Eruption of Krakatoa*, Royal Society Report of the Krakatoa Committee, 1888.)

The sun for the last three days rises in a splendid green when he is visible; about 10° above the horizon. As he advances he assumes a beautiful blue, and as he comes further on looks a brilliant blue, resembling burning sulfur . . . even at the zenith, the light is blue, varying from pale blue to a light blue later on, somewhat similar to moonlight. . . . Then as he declines, the sun assumes the same changes, but vice versa.

The eruption appears to have lowered world temperatures for two to three years by as much as one-half degree Celsius below normal (Figure 141). Some people have questioned the data, however, as most weather observations in the 1880s were made in Europe, which indeed was colder, but South America might have been warmer, and thus the true average world effects were not well known.

Harry Wexler, a former chief scientist of the U.S. Weather Bureau, was a strong proponent of the idea that volcanic eruptions affect climate. In 1952 he suggested that the warming trend of world temperatures since 1900 might be caused by the lack of major volcanic eruptions in the first half of the twentieth century. By contrast, he blamed the colder decades closing the nineteenth century on a series of major eruptions: Krakatau in Indonesia, 1883; Tarawera in New Zealand, 1886; Bandai-san in Japan, 1888; and Bogoslof in Alaska, 1890.

H. H. Lamb, a British climatologist, also has been a champion of the idea that volcanic activity affects climate. He compiled a detailed list of volcanic eruptions since A.D. 1500 and computed a "dust veil index" based on the apparent amount of volcanic debris scattered into the atmosphere. He concluded that there has been a definite relationship between world climatic trends and large volcanic eruptions.

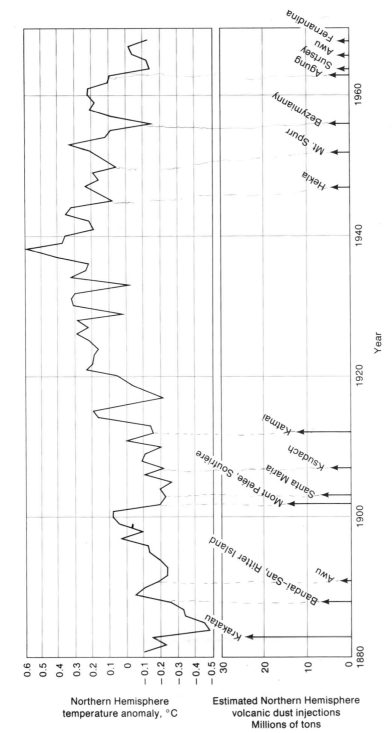

141 A graph of the average temperature variations in the Northern Hemisphere and the major volcanic eruptions that injected dust into the stratosphere. (Adapted from Robert Oliver, *Journal of Applied Meteorology* 15 [1976]: 934.)

Northern Hemisphere temperature anomaly, °C

Estimated Northern Hemisphere volcanic dust injections Millions of tons

Year

In addition, J. P. Kennett and R. C. Thunell, working with cores from deep-sea-drilling projects, concluded that the amount of volcanic ash in seafloor sediments increased about 2 million years ago and has stayed high since then. The last 2 million years coincide with the Pleistocene ice ages, and Kennett and Thunell concluded that the extra volcanism and extra cold are not just a coincidence. But even this evidence is not conclusive; some scientists point out that not enough sea cores of the older sediments have been collected to make a valid comparison.

If volcanic gas and dust do alter weather and climate, the effect probably operates in the stratosphere (above 15 kilometers), where the layer of haze hovers for a long time, since there are no clouds and rain to wash it away quickly (Figure 142). Meteorologists have identified a long-lasting stratospheric aerosol layer at heights of 15 to 30 kilometers that seems to be composed of a thin haze of small particles or droplets, smaller than 1 micron in diameter. These particles are made up of various materials, including sea salt, silicate dust, and sulfuric acid. They probably originate from several sources—sea spray, dust storms, volcanic eruptions, forest fires, industrial smokestacks, and so on. The density of the aerosol layer changes over periods of months to years. It can increase suddenly with an injection of new aerosol from a volcanic eruption, but it takes several years to decrease to normal (Figure 143).

This layer of haze in the stratosphere intercepts the incoming sunlight, heating the stratosphere and cooling the lower atmosphere as well as the Earth's surface. Measurements by the National Ocean and Atmospheric Administration's Observatory on Mauna Loa show decreases in solar transmission through the atmosphere that occurred just after the 1963 eruption of Agung Volcano on the Indonesian island of Bali, the 1982 eruption of El Chichón Volcano in Mexico, and the 1991 eruption of Pinatubo Volcano in the Philippines (Figure 143). These three volcanoes produced high eruption clouds that reached well into the stratosphere and were located in low latitudes. In addition, the El Chichón and Pinatubo eruptions released unusually large amounts of sulfur dioxide, which oxidizes to sulfuric acid aerosol in the atmosphere.

As can be seen in Figure 143, there was no apparent atmospheric effect in Hawaii from the Mount St. Helens eruption. This is attributed to three factors: except for the very beginning of the May 18 eruption, the cloud barely reached the bottom of the stratosphere; the location of

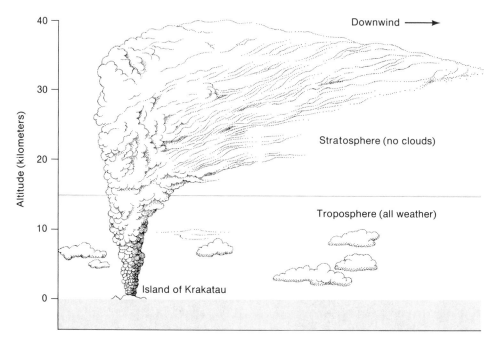

142   The Krakatau eruption cloud of 1883 was injected high into the stratosphere, where no rain clouds exist to wash the debris back to Earth. Once there, the fine dust stays in suspension for months to years. (Adapted from Richard Hazlett, 1978.)

Mount St. Helens is far from the Earth's equator; and the emission of sulfur dioxide was relatively low for the size of the eruption.

El Chichón Volcano in southern Mexico, dormant since prehistoric times and thought to be extinct, blasted into action in late March 1982. In one climactic week, three major explosions lofted large ash and gas clouds to heights of 20 to 25 kilometers. Two to three thousand people in villages near El Chichón were killed by pyroclastic flows that accompanied the eruption.

The Mount St. Helens eruption was larger in terms of volume of rock involved, but El Chichón's atmospheric effects were far greater. Besides fine ash particles, El Chichón injected an enormous amount of sulfur dioxide into the stratosphere. This gas soon oxidized, picked up water vapor, and formed tiny droplets of sulfuric acid. The acid aerosol and fine ash were carried westward by high-speed stratospheric winds, circling the Earth within a month.

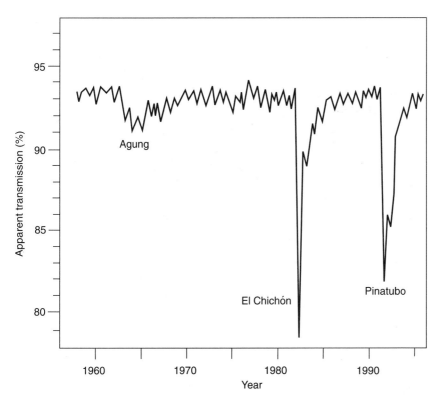

143   A graph showing the amount of direct solar radiation arriving at the Mauna
Loa Atmospheric Observatory on clear days versus time. Stratospheric haze from
explosive volcanic eruptions of Agung Volcano, Bali, Indonesia, El Chichón
Volcano, Mexico, and Mount Pinatubo, Philippines, reduced direct solar radiation
by scattering the incoming sunlight. The last two eruptions involved magma rich in
sulfur that produced haze with large amounts of sulfuric-acid aerosol. Sulfuric-acid
droplets in stratospheric haze are particularly effective in scattering solar radiation.
The apparent transmission line is smoothed from monthly averages measured at
3415 meters on the north flank of Mauna Loa Volcano, Hawaii. (Graph courtesy of
the Mauna Loa Observatory, NOAA, CMDL)

Measured by LIDAR in Hawaii, the El Chichón stratospheric haze
was 100 times more opaque than the Mount St. Helens cloud had been.
(LIDAR is a technique of aiming a laser beam into the clear night sky
and measuring the backscatter of laser light that is reflected from fine
particles or aerosol droplets suspended in the atmosphere.) It took nearly
a year for the El Chichón stratospheric cloud to attain its maximum

effect over Europe, and it was not until 1985 that solar transmission measurements returned to preeruption levels (Figure 143).

The most convincing evidence of the connection between volcanoes and world climate came in June 1991, with the huge eruption of Pinatubo Volcano in the Philippines (Chapter 18). Satellite images showed that the climactic June 15 explosions produced a mushroom cloud of ash 400 kilometers wide and 34 kilometers high. The main explosive jetting of the ash eruption lasted for three hours, and the overall magnitude of the eruption was ten times larger than the 1980 eruption of Mount St. Helens. The enormous cloud of ash and gas that spewed from Pinatubo reached high into the stratosphere, above the reach of clouds and rainfall that could wash it out.

The entrapped gases that caused Pinatubo's explosion were rich in sulfur dioxide. Oxidized in the atmosphere to small aerosol droplets of sulfuric acid, this haze completed its circle around the Earth in three weeks and then slowly spread over the sphere during the next year. The aerosol cloud reached its maximum worldwide distribution by late 1992, reducing the amount of sunlight reaching the Earth's surface during 1992–1993. The aerosol slowly decreased to more normal levels by 1995. Average world temperatures dropped about 0.4 degrees Celsius during 1992–1993, bucking the trend of global warming during the past two decades.

The Earth has been in a warming trend since 1980. Some meteorologists blame this on the increasing amount of carbon dioxide in the atmosphere, caused by the burning of fossil fuels and the destruction of rain forests that consume carbon dioxide. Carbon dioxide in the atmosphere has a "greenhouse effect": it is transparent to light but more opaque to infrared radiation. Although the warming of the Earth by sunlight during the day is not affected, the cooling of the Earth at night by infrared radiation into space is impeded. The carbon dioxide acts like the windows in a greenhouse: light can come in, but heat cannot escape. The net effect is a global warming of the Earth's surface. Without this warming greenhouse effect, the cooling influence of the volcanic clouds from El Chichón and Pinatubo might have been much more pronounced.

The question remains: To what extent do volcanic eruptions affect our weather and climate? Evidence shows that eruptions injecting large amounts of fine ash and, even more importantly, great quantities of sulfur dioxide gas into the stratosphere do significantly affect the solar radiation reaching the Earth's surface. Major explosive eruptions, particularly in equatorial regions where their ash and gas clouds may affect both hemispheres, appear to be capable of lowering average global temperature by as much as 1 degree Celsius. An extremely large event like the Yellowstone eruption 600,000 years ago, or a series of major explosive eruptions that prolong the stratospheric haze layer for decades or centuries, would have a much greater cooling effect.

Volcanic eruptions are only one factor among many that affect weather and climate; the normal interaction of the atmosphere, oceans, and land surface is extremely complex. Add to this the disruption by both man-made and natural causes, and the intricacy of the problem becomes apparent. It is a joint task for geologists, meteorologists, and climatologists to understand and quantify these effects. Such understanding is of immense importance to a world that depends on a benign climate for its very existence.

Sources

Angell, J. K. "Comparison of Stratospheric Warming Following Agung, El Chichón, and Pinatubo Volcanic Eruptions." *Geophysical Research Letters* 20 (1993).

Defoor, T. F., E. Robinson, and S. Ryan. "Early LIDAR Observations of the June 1991 Pinatubo Eruption Plume at Mauna Loa Observatory, Hawaii." *Geophysical Research Letters* 19 (1992).

Lamb, H. H. *Climate: Present, Past, and Future*. London: Methuen, 1972.

McCormick, M. P., L. W. Thomason, and C. R. Trepte. "Atmospheric Effects of the Mt. Pinatubo Eruption." *Nature*, February 2, 1995.

Rampino, M., et al. "Volcanic Winters." *Annual Review of Earth and Planetary Sciences* 16 (1988).

Robock, A., and J. Mao. "The Volcanic Signal in Surface Temperature Records." *Journal of Climate* 8 (1995).

Self, S., R. Turco, et al. *Volcanism and Climate Change*. American Geophysical Union, Special Report, May 1992.

Forecasting
Volcanic Eruptions

144 A seismogram of small earthquakes related to volcanic activity in Hawaii. (Photograph by the U.S. Geological Survey.)

Nature cannot be commanded except by being obeyed.

—Sir Francis Bacon (1561–1626)

F orecasting the time, place, and character of volcanic eruptions is one of the major goals of volcanology. We prefer the word *forecast* to *prediction* because the science of weather forecasting has established that forecasts are probabilistic; that is, they are not precise, though the hope is that they are more accurate than statistical averages. For example, suppose that rainfall records in Hilo, Hawaii, show that on average it rains during six out of ten days throughout the year. If a weatherman forecasts a 60 percent chance of rain tomorrow in Hilo, he is not making a very adventurous statement. But if the wind patterns and satellite photos indicate that the chances of rain tomorrow are greater or less than average, then the forecast may say a 90 percent chance of rain or a 10 percent chance of rain.

These would be valid forecasts based on information other than historical statistics; their worth can be evaluated in hindsight by comparison with random guesses. Weather forecasters avoid the word *prediction* because it sounds so precise and specific. Even though current weather forecasts are far from exact, they are extremely valuable. They do much better than random guesses, and their batting average is improving.

The current goal in forecasting volcanic eruptions is to provide the best forecasts possible, based on the geologic history of the volcano under study, as well as on the day-to-day vital signs of the volcano in terms of earthquakes, surface deformation, temperature, gas emissions, and other measurements. The statistics of past eruptions are of great importance, both as a basis for finding the average probability of eruption and as a means of deciphering some pattern in eruption habits. At least three patterns have been recognized, even though the periods of repose between eruptions may vary greatly.

One pattern is completely random. That is, no matter how long the repose period has been, the average chance for an eruption next month remains the same. This is like cutting cards to get an ace; no matter how many times you fail, the chance in the next cut is exactly the same: four out of fifty-two. Mauna Loa Volcano in Hawaii appears to operate in this random manner; no matter how long or short the repose between eruptions, the average chance for a new eruption during the next month remains the same—2 percent (Figure 145).

Hekla Volcano in Iceland shows quite a different pattern of time intervals between eruptions. At Hekla the average probability of an eruption increases with time. This is like cutting for aces and discarding the cut card each time you fail, thereby increasing the chance of getting an ace with each new cut.

Just the opposite holds for volcanoes like Kilauea in Hawaii, where a group of eruptions cluster together in time (Figure 146). In this situation the probability of an eruption decreases with time. There is no easy analogy for this in card cutting.

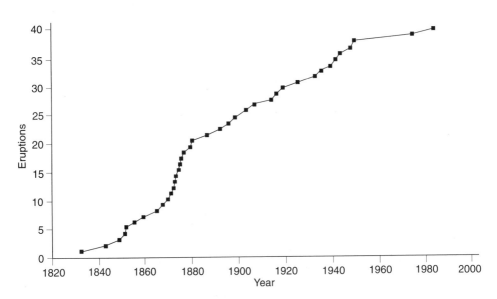

145   The cumulative number of historical eruptions of Mauna Loa Volcano in Hawaii plotted against their onset dates. The average recurrence time is four years. However, the frequent eruptions between 1870 and 1880 and the long intervals since the eruption of 1950 suggest distinct clusters of nonrandom recurrence times.

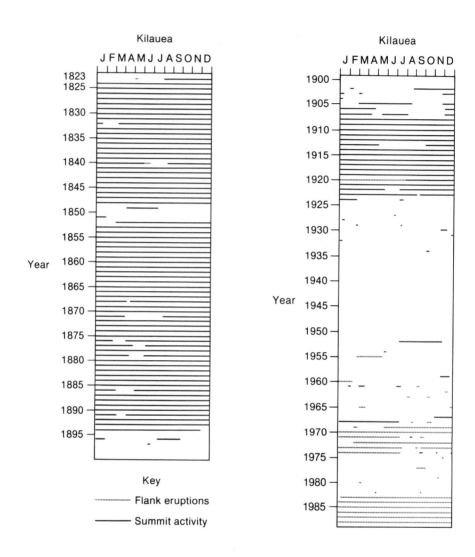

146   Eruptions of Kilauea Volcano in Hawaii since 1823 show a pattern of long periods of continuing eruptions or groups of eruptions interspersed with long periods of repose. The periods from 1823 to 1894 and from 1906 to 1924 were characterized by nearly continuous lava lake activity in the summit crater. (After Gordon Macdonald and Douglass Hubbard, *Volcanoes of the National Park of Hawaii*, Hawaii Natural History Association, 1982.)

TABLE 7
Number versus size of explosive eruptions

VEI	Volume	Number	Adjective	Example
1	10^4–10^6 m^3	100/yr	Small	Stromboli 1996
2	10^6–10^7 m^3	15/yr	Moderate	Unzen 1991
3	10^7–10^8 m^3	2–3/yr	Moderate–large	Ruiz 1985
4	10^8–10^9 m^3	1 per 2 yrs	Large	Rabaul 1994
5	1–10 km^3	1 per 10 yrs	Very large	Mount St. Helens 1980
6	10–100 km^3	1 per 40 yrs	Huge	Pinatubo 1991
7	100–1000 km^3	1 per 200 yrs	Colossal	Tambora 1815
8	1000–10,000 km^3	1 per 50,000 yrs	Humongous	Yellowstone 630,000 B.C.
9	>10,000 km^3	Not found	Fortunately	None

Each of these time patterns in eruptive habit is important as a basis for forecasting future activity. Unfortunately, for the statistics to have any meaning, the number of eruptions must exceed ten or twenty, and only a small fraction of the world's volcanoes are active enough or have been studied long enough to establish their patterns. It is possible, however, to estimate how often volcanic eruptions of differing sizes will occur worldwide. Of course, the number of small eruptions greatly exceeds that of large ones, but this number can be quantified. Table 7, based on statistics of volcanic eruptions compiled by the Smithsonian Institution and the U.S. Geological Survey, indicates the number of explosive eruptions of a particular magnitude or larger that have taken place during the past two hundred years, plus the extremely large explosive eruptions that apparently have occurred during the past million years. Extrapolating this table into the future gives a rough estimate of how often variously sized eruptions will take place in the years and centuries to come.

Dormant volcanoes are by no means dead, and by studying them with sensitive instruments it is possible to monitor their vital signs

Magnitude 3 earthquake

147   A seismograph recording earthquakes at the Hawaiian Volcano Observatory. The drum rotates once every 15 minutes, slowly enough that an entire day can fit on one record sheet. Most of the microearthquakes on this record are too small to be felt. However, the larger quake, whose seismogram is just to the left and below center, was about magnitude 3, large enough to be felt in a local area close to its source. (Photograph by the U.S. Geological Survey.)

through periods of repose and awakening. One of the most important of these signs is the number of earthquakes and their locations.

Both earthquakes and volcanoes occur along plate margins, where many of the earthquakes are considered to be related to the slow slipping and breaking of the plates' moving edges. Some earthquakes, how-

ever, appear to be more directly related to volcanic processes. In Hawaii
and other areas of hot-spot volcanoes, earthquakes accompany volcanism, even though the plate margins may be thousands of miles away
(Figures 144 and 147). In fact, Hawaii's Kilauea Volcano is one of the
most seismically active areas on Earth. Ten or more microearthquakes
occur beneath the volcano on an average day, and there is an earthquake large enough to be felt about once a week.

During the immediate prelude to an eruption when new magma
conduits are opening underground, hundreds to thousands of microearthquakes shake the seismometers, with several felt shocks generally accompanying these earthquake swarms. At volcanoes like Mount
St. Helens that are related to subduction zones, the number of microearthquakes during dormant periods is usually only a few per month,
but in swarms preceding an eruption the count of small quakes generally increases to several hundred per day.

There is probably more than one cause for these volcanic earthquakes, including increasing topographic load, underground temperature changes, moving magma, and gas explosions. The slipping and
cracking of rocks underground to adjust to the growing weight of a huge
volcano is only indirectly related to volcanic activity, but the other
causes, particularly the movement of magma and the formation of
cracks through which it can move, are closely connected to active volcanic processes.

Studies by Bernard Chouet of the U.S. Geological Survey indicate
that swarms of long-period (LP) earthquakes occurring at shallow
depths beneath a volcano seem to be useful indicators of an impending
eruption. These precursors may occur over periods as short as days to
weeks before an eruption. Swarms of such LP earthquakes have clearly
preceded recent eruptions at Redoubt Volcano in Alaska, Pinatubo Volcano in the Philippines, and Galeras Volcano in Colombia.

The number or size of volcanic earthquakes, particularly those
related to the conduits through which magma erupts to the surface,
usually increases before an eruption. This relationship is not infallible,
however. In a study of seventy-one earthquake swarms and volcanic
eruptions, 58 percent showed an increase in earthquake activity before
an eruption, 38 percent showed an increase without an eruption, and
4 percent had no apparent increase in earthquake activity before an
eruption. This study by David Harlow, who was then a student at

Dartmouth College, dates back to 1970. Since then, there have been many earthquake swarms and eruptions, and a 1996 study by John Benoit and Stephen McNutt at the University of Alaska shows remarkably similar results. Their survey of 327 earthquake swarms beneath potentially active volcanoes between 1979 and 1989 found that 191 of the swarms were followed by eruptions (58 percent) and 136 (42 percent) were not. The swarms not followed by eruptions had a shorter average duration (3.5 days) than did the precursory swarms (8 days).

Because the background count of microearthquakes in volcanic areas is highly variable, only a large change in their number—by as much as a factor of 100—seems to be significant. Russian scientists in Kamchatka now believe that large increases in the total energy released by volcanic earthquakes is more important than increases in their number. Barry Voight, at Pennsylvania State University, argues persuasively that it is the acceleration of the earthquake count or energy that is the key figure. The time between the beginning of an earthquake swarm and the actual eruption varies from months to hours, but even so, the number, size, and location of earthquakes on active volcanoes are an important index of forthcoming activity.

Volcanic tremor is a unique kind of seismic activity associated with volcanoes. It consists of more-or-less continuous ground vibration with a frequency of 0.5 to 10 cycles per second—a very low hum detectable by seismographs (Figure 148). Its source is not clear; various studies relate it to the formation of gas bubbles or the irregular flow of magma, which creates vibrations like the noise of water running in poorly designed pipes. Whatever its source, it is nearly always present during a volcanic eruption and often begins before the actual surface outbreak. Not all periods of volcanic tremor are followed by an eruption, but in Hawaii, high-amplitude volcanic tremor is often the best indication that an eruption has begun or is about to begin within a few minutes or hours. A "tremor alarm" ringing in their houses alerts the observatory staff to check their instruments.

Slight changes in the angle of slope or distance between survey points on the summits and flanks of active volcanoes provide another important method of diagnosing internal changes taking place. Several techniques are involved, including conventional leveling and the determination of distances with reflected laser beams. Tiltmeters, which can detect changes in slope smaller than 1 part per million, are also used.

1 minute

148   Volcanic tremor recorded on a seismograph in Hawaii during the eruption of Kilauea in 1977. This low-frequency vibration of the ground at about 3 to 5 cycles per second is related to the movement of magma through underground conduits. The tremor starts before the breakout of eruptions and is a sign that an eruption is likely. (Photograph by James Griggs, U.S. Geological Survey.)

An angular change in slope of 1 part per million (1 microradian) is equivalent to lifting the end of a rigid board 1 kilometer long by only 1 millimeter. A technique called GPS (Global Positioning System) uses satellites and portable receivers to determine the relative position of the ground stations to accuracies of a few millimeters (Figure 149). Radio code and time signals broadcast by the satellites arrive at ground-surface survey stations in a unique pattern. A comparison by computer processing of the different patterns recorded at two survey stations determines their relative location. GPS receivers that monitor changes in position almost continuously have recently been installed at Augustine Volcano in Alaska and at Kilauea Volcano in Hawaii.

Both elevation and distance changes can be detected by GPS measurements. The receivers can be many kilometers apart and need to

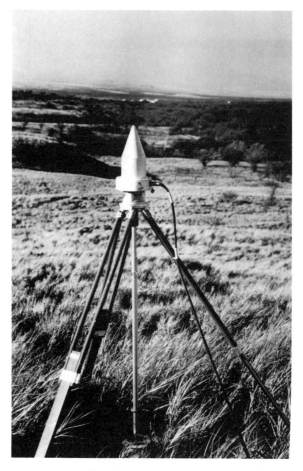

149 A GPS (Global Positioning System) antenna is set up over a survey benchmark to receive radio time and code signals from passing satellites. Elevations and distances between two or more receivers operating simultaneously can be determined to accuracies of better than 1 part per million.

"see" the satellites but not one another. Reasonably priced GPS instruments are now available, and important observations can be made in a few hours by a small group of surveyors. Recent GPS surveys in Hawaii revealed that the southeast flanks of both Kilauea and Mauna Loa are slowly moving a few centimeters per year toward the southeast (Figure 150).

150 Arrows show the direction and velocity of the slow movement of the southeast flank of Kilauea Volcano towards the sea. This movement has been measured using the satellite Global Positioning System (GPS) from 1990 through 1995. The ellipses show the uncertainty of the measurements and the point labeled BLDR is the reference station. (Data and map from Mike Lisowski, Hawaiian Volcano Observatory, U.S. Geological Survey.)

The latest technique to "see" ground-surface deformation on volcanoes is called *satellite interferometry*. Digital radar images of the topography, taken at different times, are subtracted from one another. If uplift or subsidence has occurred, the changes appear as interference rings or bands of iridescent colors, like the colored rings or bands seen in the reflected light from a thin oil slick on water. If there has been no deformation, no interference bands will appear. Although still in an experimental phase, this technique seems to have great promise. Satellite radar surveys at periodic intervals could monitor many potentially dangerous volcanoes and also provide basic research data on their internal structure and dynamics.

Repeated at frequent intervals, any of these surveying techniques may reveal the tiny deformations of the volcano's surface caused by changes in magmatic pressure or volume inside the volcano. *Deformation monitoring*, the collective name for all these surveying techniques, has become the second most important means of forecasting volcanic eruptions, preceded only by the monitoring of earthquakes.

The slow inflation of Kilauea Volcano in the months before an eruption, and the sudden collapse over a period of hours or days during an eruption (described in Chapter 7) have been revealed by deformation measurements in the past forty years (Figure 151). The slow, more-or-less continuous addition of magma into the shallow chamber beneath Kilauea's summit causes the summit to swell upward as much as 2 meters over a diameter of 10 kilometers. This gentle bulge causes outward tilts and increasing distances between survey stations, similar to the changing relationship of spots on the surface of an inflating balloon.

Intermittent eruptions of Kilauea between 1955 and 1986 removed molten lava from the magma chamber faster than it was replenished, causing summit deflation, inward tilt, and contracting distances between survey stations. Since 1986, Kilauea has been erupting almost continuously, and magma is being replenished nearly as fast as it is erupted. During the intermittent eruptions, deformation measurements provided a kind of magma barometer that in turn could be used to help forecast eruptions. If there were an exact level of reinflation that triggered a new eruption, the question of when an eruption will occur would be solved.

But nature is not that simple. On Kilauea, which has been tracked through more deformation cycles than any other single volcano, the degree of tilt at the Hawaiian Volcano Observatory just before an eruption has always been relatively high, but has not shown any specific critical angle. The reason for this may be that an increase in either pressure or volume in the shallow magma chamber beneath Kilauea can cause uplift and outward tilt of the summit area. Slow increases in volume may not cause equal increases in pressure, and it is increasing pressure that causes the hard rocks surrounding the magma chamber to split apart. Magma moves into these newly formed underground cracks, and the summit subsides rapidly. The magma may halt underground in these new fractures as a shallow intrusion, or it may break to the surface in an eruption.

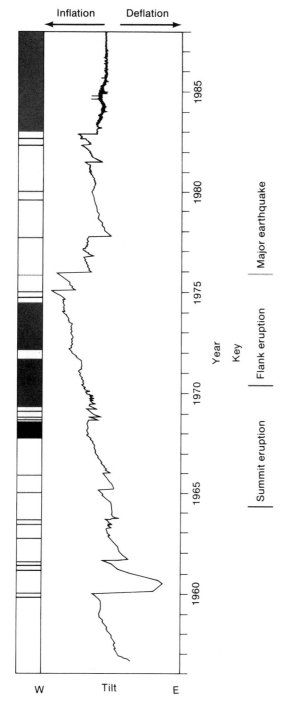

151 The 30-year tilt record of Kilauea Volcano in Hawaii shows the ups and downs related to eruptions. For months to years before an eruption, the summit of Kilauea inflates. When it reaches a level that exceeds the strength of the rocks surrounding the shallow magma chamber, the volcano erupts. Some eruptions rapidly remove large volumes of magma from the summit chamber, and the volcano deflates over a period of days. A high level of inflation indicates a higher probability of eruption, but since the strength of the rocks surrounding the magma chamber varies from one eruption to another, no exact level of tilt can be used to predict the next eruption. (Data from the U.S. Geological Survey.)

Deformation measurements on several potentially explosive volcanoes on subducting plate margins are now being made, but it is not yet clear whether they will be as useful as those on basaltic shield volcanoes. One problem with explosive volcanoes is their long repose time between eruptions. It could take several hundred years to learn as much about the deformation at Vesuvius as has already been learned at Kilauea in the last forty years. Nevertheless, monitoring both the earthquake swarm and the large surface bulge that took place for nearly two months before the major eruption of Mount St. Helens in 1980 provided important insights into understanding what was going on beneath the volcano during that period (see Chapter 5).

Changes in the Earth's magnetic and electrical fields near volcanoes, which relate to the state of volcanic activity, have been observed in Japan, New Zealand, Kamchatka, and Hawaii. Although the techniques used for these observations are not as well established as seismic and deformation measurements, they look promising.

Temperature changes at steam vents and warm springs on volcanoes would seem to be an obvious index for forecasting, but rainfall and changes in groundwater circulation often cause large fluctuations in temperature not related to volcanic activity. In one case, however, a $12°C$ temperature rise of the crater lake at Taal Volcano in the Philippines clearly signaled its 1965 eruption (Figure 152). Geochemical changes in the volume and composition of volcanic gases also are useful indicators of hidden changes beneath active volcanoes. New steam vents formed at Askja Volcano in Iceland two weeks before its eruption in 1961. Scientists in Japan, Kamchatka, and the United States all have reported an increase in sulfur gases relative to chloride gases at volcanic steam vents in the years or days before some eruptions. Changes in the percentages of hydrogen, helium, and radon in volcanic gases are also being studied as possible signals of changing volcanic activity.

No single technique appears to be the master key to forecasting volcanic eruptions. Each volcano is unique, and the case history of one cannot always be used to diagnose the symptoms of another. Even so, useful though not precise forecasting is currently being practiced on volcanoes in Japan, Indonesia, Iceland, the Philippines, Kamchatka, Hawaii, and Alaska. The success in forecasting the 1975 and 1984 eruptions of Mauna Loa Volcano illustrates the present state of the art.

152   Changes in the temperature of the crater lake on Taal Volcano in the Philippines sometimes warn of forthcoming eruptions. In 1965 the temperature began to rise well above its background level of 33°C in July; the volcano erupted in September. The 1966 eruption was preceded by a much less obvious temperature rise, and the 1967 eruption could not have been forecast on the basis of temperature alone. (Data from A. Alcaraz, Philippine Commission on Volcanology.)

Mauna Loa Volcano in Hawaii erupted on July 5 and 6, 1975, after twenty-five years of repose—its longest sleep since records began in the mid-nineteenth century. Thirty million cubic meters of lava poured out in that brief but intense summit eruption. The monitoring of earthquakes and deformation of Mauna Loa was increased during and after the eruption, and scientists waited with interest to find what these observations would reveal while Mauna Loa rested and prepared itself for the next eruption. It was a ten-year wait, and Figure 153 sums up the data.

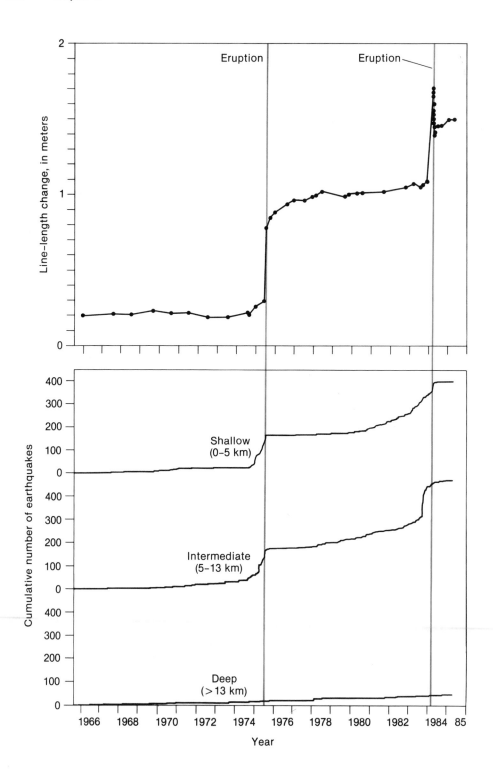

153   *(Facing page)* Earthquakes and deformation are precursors to eruptions of
Mauna Loa in Hawaii. Shallow and intermediate-depth quakes beneath the
volcano, most of them too small to be felt, increase for one to two years before an
eruption. The 3-kilometer-wide caldera widens mainly during eruptions (each dot
is a separate survey). (After J. P. Lockwood et al., U.S. Geological Survey
Professional Paper 1350, 1987, pp. 548–549.)

    Small earthquakes of shallow and intermediate depth had shown an
increase for one to two years before both the 1975 and 1984 eruptions,
but the number of earthquakes deeper than 13 kilometers showed no
change. The sharp increase in intermediate-depth earthquakes in late
1983 was caused by numerous aftershocks that followed a magnitude
6.6 earthquake that had occurred beneath the east flank of Mauna Loa
in November of that year.

    Deformation measurements of the summit showed a 75-millimeter
increase in the length of a 3-kilometer-long survey line across the caldera in the year before the 1975 eruption, as well as a similar increase
before the 1984 activity. At the onset of both eruptions, the summit
caldera widened suddenly by about 500 millimeters, caused by the
opening of magma-filled cracks that split the caldera and fed the
eruptions.

    Following the start of the eruption in 1984, the width of the caldera
contracted by 300 millimeters, because the Northeast Rift Zone vents
at 2900 meters drained part of the summit magma reservoir and caused
the summit to subside and contract. After the three-week-long 1984
eruption of 220 million cubic meters of lava (Figure 154), the summit
began to reinflate and widen again. By 1997 the summit inflation and
stretching had recovered about 50 percent of the subsidence and contraction that accompanied the flank eruption. Mauna Loa's long repose
periods since 1950 suggest that the statistical probability of another
eruption occurring next month has dropped to as low as 0.5 percent
(1 chance in 200). The lack of any increase in shallow earthquakes
beneath the summit of Mauna Loa at this time (1997) supports this
forecast.

    The precursors for the 1975 and 1984 Mauna Loa eruptions were
on two time scales: the earthquake increases that began about a year
before and the volcanic tremor that preceded the outbreaks by only one

Key

0   5   10 km

Contour interval 1000 meters   ┼┼┼┼┼┼┼┼ Eruptive vents

Lava flows

154   Fissure vents and lava flows of the 1984 Mauna Loa eruption. After the initial outbreak at the summit, eruptive fissures opened progressively down the northeast rift. The fractures at the 2900-meter elevation opened 15 hours after the eruption began and remained the principal vents of the 3-week-long eruption. Flow 1 reached its maximum length in 5 days and then diverted itself upstream into flow 1A. Flow 1B was formed by similar channel clogging of flow 1A on April 5, and Hilo was no longer threatened by the eruption. (After J. P. Lockwood et al., U.S. Geological Survey Professional Paper 1350, 1987, p. 561.)

to two hours. Some precursors a few days or weeks before a Mauna Loa eruption would be most useful to discover.

The Russians may be doing better. The huge eruption of Tolbachik Volcano in Kamchatka in 1975 was preceded by a major earthquake swarm. Volcanologist P. I. Tokarev's published forecast that an eruption was imminent in the Tolbachik region during the next week allowed Russian television crews to be on hand for the birth of a new volcano on July 6, just two days after the forecast was issued (Figure 155).

155   The great Tolbachik eruption in Kamchatka in 1975 was accurately forecast on the basis of the increasing energy of local earthquakes. The total volume of lava and pyroclastic material erupted was nearly 2 cubic kilometers, and the ash cloud reached 14 kilometers high. (Photograph by N. P. Smelov, Institute of Volcanology, Kamchatka.)

The classic success story in volcano forecasting is the evacuation of the area around Pinatubo Volcano in the Philippines just before its catastrophic eruption on June 15, 1991. That forecast, based on increasing seismicity and a prehistoric record of large explosive eruptions, saved tens of thousands of lives, as described in the next chapter.

Sources

Alaska Volcano Observatory. *BiMonthly Reports*. Vols. 1–9. University of Alaska and U.S. Geological Survey, 1989–1997.

Benoit, J. P., and S. R. McNutt. "Global Volcanic Earthquake Swarm Database and Preliminary Analysis of Volcano Earthquake Swarm Duration." *Annali di geofisica* 39 (1996).

Chouet, Bernard. "Long-Period Volcano Seismicity: Its Source and Use in Eruption Forecasting." *Nature*, March 28, 1996.

Decker, R. W. *Forecasting Volcanic Eruptions*. Palo Alto, CA: Annual Reviews of Earth and Planetary Sciences, 1986.

Decker, R. W. "How Often Does a Minoan Eruption Occur?" *Thera and the Aegean World III*. Vol. 2 (Earth Sciences) London: Thera Foundation, 1990.

McGuire, B., C. R. J. Kilburn, and J. Murry. *Monitoring Active Volcanoes*. London: UCL Press, 1995.

McNutt, S. R. "Volcanic Tremor." In vol. 4 of *Encyclopedia of Earth Science Systems*. San Diego: Academic Press, 1992.

Rhodes, J. M., and J. P. Lockwood, eds. *Mauna Loa Revealed*. American Geophysical Union Monograph no. 92, 1995.

Sabins, F. F. *Remote Sensing*. 3rd ed. New York: Freeman, 1997.

Swanson, Donald, et al. "Forecasts and Predictions of Eruptive Activity at Mount St. Helens, USA." *Journal of Geodynamics* 3 (1985).

Voight, B. "A Method for Prediction of Volcanic Eruptions." *Nature*, 1988.

Reducing
Volcanic Risk

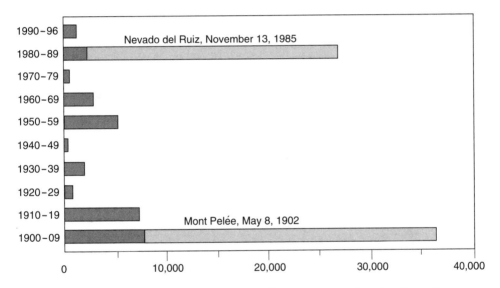

156 Volcano fatalities by decade in the twentieth century through 1996. Most of the deaths occurred in just two events indicated in light gray—the pyroclastic flow from Mont Pelée on the Caribbean island of Martinique and the mudflow from Nevado del Ruiz in Colombia. (After Robert Tilling, U.S. Geological Survey.)

T he effort to forecast volcanic eruptions is not just an exercise in scientific curiosity; the goal is to be able to reduce volcanic risk in an increasingly overpopulated world (Figure 156). With each decade the number of people living close to—or on—dangerous volcanoes increases dramatically. Some are drawn by the wild, scenic beauty, others by the rich volcanic soils or the population pressure on crowded islands.

The threat of a volcanic eruption in such an area calls for fast response in many forms, including continuous monitoring of the awakening volcano's vital signs; assessment of potential hazards by studying its eruptive history and mapping the deposits around it; and—one of the most important components—good communication with local officials, the public, and news organizations. Residents generally wish to downplay the dangers; newspapers often exaggerate them; and scientists are squeezed in between.

In dealing with nature, especially in a form as capricious as a volcano, things seldom go as planned. One instance in which forecasting, contingency planning, and action all came together to achieve a remarkable success was the 1991 eruption of Pinatubo Volcano in the Philippines. The mountain stirred to life in March of that year with swarms of small earthquakes that were felt by local residents, followed a couple of weeks later by the first steam explosions near the summit. It had been 500 years since Pinatubo last erupted, and most people living in the region did not even know that it was a volcano. It did not have the classic shape of a volcano; its 1700-meter summit was eroded into a ragged ridge and its steeper slopes were covered with jungle.

Fortunately, scientists from the Philippine Institute of Volcanology were well equipped and ready to take action. They immediately started installing seismographs and other monitoring equipment and were soon joined by colleagues from the U.S. Geological Survey. Together they made a rapid reconnaissance of the nature and distribution of prehistoric eruption deposits from the volcano, and the results were bad news: Pinatubo had a dangerous past. By radiocarbon dating of charcoal fragments in old volcanic deposits, the geologists determined that there had been major explosive eruptions about 500, 3000 and 5500 years ago, and that these eruptions had poured out enormous pyroclastic flows, thick ashfalls, and great aprons of lahar deposits around the mountain. Those prehistoric deposits on the east side of Pinatubo were flat and reasonably level—a perfect surface for the giant runways of the U.S. Air Force's Clark Air Base that was built upon them.

Pinatubo's awakening progressed rapidly. The earthquake energy kept increasing, and the location of the earthquakes migrated from a zone beneath the north side of the volcano to a more shallow zone closer to the summit. During May, the emissions of sulfur dioxide gas rose dramatically to more than 5000 tons per day before dropping again in early June. Small ash explosions began on June 3, and a lava dome pushed out on the northwest side of the summit on June 7, producing a cloud of steam and ash 7 kilometers high.

The Philippine Institute of Volcanology devised a five-level alert scheme in May when it became evident that with Pinatubo's past history, the growing unrest posed a grave danger to the population near the mountain. The Institute and the U.S. Geological Survey launched a massive program to communicate the severity of the situation to Philippine officials and civic leaders, and at the same time to educate them about the hazards that could occur.

One device that was particularly effective was the use of a video entitled "Understanding Volcanic Hazards" produced by Maurice and Katia Krafft (Figure 157) for the International Association of Volcanology, showing vivid examples of pyroclastic flows, lahars, and ashfalls to officials who had never seen such phenomena in action. The film was copied and shown to government officials, countless civic groups, and as many individuals as could be reached; it was credited with convincing officials and public alike that the threat was real.

With good management, cooperation with local governing officials, and some luck, by June 14, the day before the climactic eruption, 60,000 people had been convinced to evacuate a zone within a radius of 30 kilometers from the summit of Pinatubo (Color Plate 17). On

157   Katia and Maurice Krafft, French volcanologists, at the Kupaianaha Lava Lake in Hawaii. They studied and photographed volcanoes all over the world, witnessing and documenting more eruptions than any other investigators. Their popular books, lectures, and videotapes greatly increased public awareness of the nature and dangers of volcanic eruptions. Ironically, along with American volcanologist Harry Glicken, they were killed by a pyroclastic flow from an eruption of Unzen Volcano in Japan in 1991.

June 10, Clark Air Base evacuated 14,500 U.S. personnel, and on June 15, the destruction at the installation was so severe that no attempt was made to renew the base contract with the Philippine government.

On June 15, huge explosions produced a mushroom cloud of ash 400 kilometers wide and 34 kilometers high. Unfortunately Typhoon Yunya arrived at the same time and made eyewitness observations almost impossible, but satellite images and recording barographs indicated that the main explosive jetting of the ash eruptions lasted for about three hours. The volume of ash was enormous, estimated at about 4 cubic kilometers. Huge pyroclastic flows poured from Pinatubo and swept as far as 16 kilometers from the summit. These hot flows of pumice and ash sterilized 400 square kilometers of land surrounding the volcano with high-temperature deposits as much as 200 meters thick, while heavy rains from the typhoon mixed with the ashfall and pyroclastic flow deposits, and stirred up thick lahars that poured down all the streams and rivers around the mountain (Figure 158).

About 300 people lost their lives, but most of them were killed when roofs collapsed under the heavy, wet, ash deposits. Without the typhoon, the toll presumably would have been smaller, but without the accurate forecast and prompt action, the death toll could have been in the tens of thousands. However, in the years since the eruption, hundreds of people have died from conditions in relocation camps and from the lahars that continue to plague the region.

The eruptions of Mount St. Helens (described in Chapter 5) illustrate the current strengths and weaknesses of volcano forecasting. Based on that volcano's geologic history, a warning was published in 1978 that destructive eruptions might occur within a few decades. After small eruptions began in March 1980, the strong, ongoing earthquake swarm and fast-growing bulge indicated to geologists monitoring Mount St. Helens that a sizable shallow intrusion of magma was being injected beneath the north flank of the mountain.

The probability that the magma would reach the surface in a significant eruption was considered quite high, but the timing and scale of the great May 18 eruption were not anticipated. Even when reviewing the events now, no specific precursors or precedents can be detected that might have called for a complete evacuation of the area that was later devastated. Nevertheless, since the danger caused by the shallow intrusion of magma was recognized, access to the area was severely restricted, and many hundreds of lives were saved.

After the great eruption, there were five smaller explosions during 1980, and the lava dome in the crater grew rapidly during several periods between 1980 and 1987. Beginning with the July 1980 explosion, all the eruptions and dome-building events at Mount St. Helens have been accurately forecast from increases in seismicity and rates of deformation. These precursors, which have occurred days to hours before

158   A map showing the major deposits from the huge eruption of Mount
Pinatubo in the Philippines. The airfall ash deposits were from the explosions of
June 12 to 15, 1991; the pyroclastic flow deposits were mainly from the climactic
eruption on June 15, 1991; and the lahar deposits continued to plague the region
every rainy season for five years or more after the eruption. (After the Pinatubo
Volcano Observatory Team, *EOS*, December 3, 1991.)

each eruptive event, have given sufficient warning to allow evacuation of the hazardous area.

Most infamous volcanoes owe their notoriety to the deaths and destruction they caused. Those eruptions that were forecast—thereby saving thousands of lives—are less well known. One such example occurred in Indonesia in 1983. Colo Volcano, on the island of Una Una, was shaken by an earthquake swarm that began on July 14, with small explosive eruptions starting on July 18. On the basis of the past behavior of Colo and similar volcanoes, geologists from the Volcanological Survey of Indonesia recommended to local government officials that the island be evacuated. The officials agreed, and all 7000 inhabitants were removed by boat. On July 23 the climactic eruption swept the island with hot pyroclastic flows. Most of the livestock and coconut plantations were destroyed. It has taken years to rebuild the island's economy, but the people survived.

Villagers who were living in Armero at the foot of Colombia's Nevado del Ruiz volcano when it erupted in 1985 (described in Chapter 10) were much less fortunate. Geologists had warned them that because of the ice cap on the high summit and a lethal history of lahars, any eruption of molten or hot fragmental rock onto the summit ice could bring disaster. But no one in Armero really knew what a lahar or mudflow was; when the volcano erupted, only a few people evacuated the town. There was a two-hour period between the main summit eruption and the time the lahar swept over Armero, but the summit of Ruiz was not visible, and no one could see the approaching catastrophe. In hindsight the tragedy might have been prevented. In reality it was not.

This seems to have been a case of a massive failure of communication. Authorities knew of the critical danger from lahars but failed to give an unambiguous warning to evacuate. The canyon above the village was not equipped with a flood-warning system—used commonly in Japan—which could have given ample warning. Also, the use of a hazards video that proved to be so effective in educating the people who were at risk from Pinatubo could have made a life-or-death difference here.

In general, volcanologists and public officials have an easier task getting public cooperation in a crisis in areas where the people have a

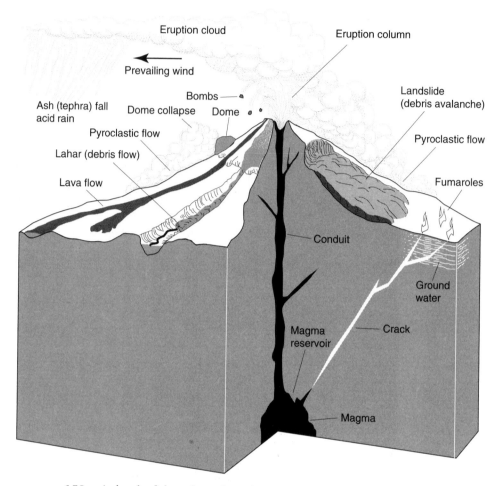

159   A sketch of the volcano hazards posed by a typical stratovolcano. Volcanic risk, the danger from the hazards to living things and property, can be reduced by public awareness of volcanic processes and products. For example, living upwind from a volcano is safer than living downwind, and living in a valley near a volcano is more dangerous than living on a hillside. (From Bobbie Myers and Steven Brantley, U.S. Geological Survey.)

memory of volcanic activity and especially where there has been education in preparedness (Figure 159).

This was the case at Rabaul in Papua New Guinea in 1994. The city of Rabaul is situated on the shore of a bay that is actually a collapsed caldera. Its location—on the Ring of Fire—is a clue that eruptions there have been explosive. Other obvious clues to the island's volcanic origins are the many volcanic cones that dot the caldera's rim. Two prominent vents, Tavurvur and Vulcan, stand across the bay from each other like sentinels guarding the entrance to the inner harbor (Figure 160).

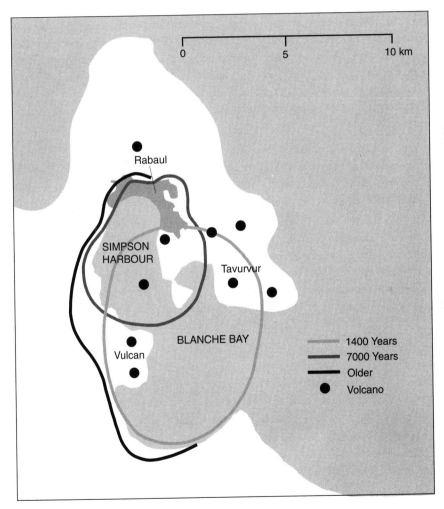

160 The sheltered harbor at Rabaul in Papua New Guinea is the result of multiple caldera collapses, the latest about 1400 years ago. Along with the benefits of a good harbor comes the danger of volcanic eruptions; the most recent occurred from both Vulcan and Tavurvur Volcanoes in 1994. (After Russell Blong and Chris McKee.)

Geologists believe that the caldera has collapsed several times in the last 10,000 years. They have found charcoal fragments in pyroclastic deposits surrounding Rabaul Caldera with radiocarbon dates of about A.D. 500 to 600 for its last huge collapse. Since that great eruption, many smaller ones have built the cones scattered along the caldera rim.

In 1937 an eruption began in the shallow, near-shore waters of the caldera and built a 230-meter-high cone, Vulcan, most of it during the first twelve hours. Hot pyroclastic flows from Vulcan killed more than 500 people. Tavurvur, 5 kilometers across the entrance of the bay, also erupted during the brief 1937 event. Rabaul Town escaped the deadly pyroclastic flows but was showered with 7 to 15 centimeters of volcanic ash.

The Rabaul Volcano Observatory (RVO), which was established before World War II but was closed during the Japanese occupation of Rabaul from 1942 to 1945, opened again in 1950. In 1973, scientists there began making some startling observations: the caldera floor was slowly bulging upward. Over the next decade they measured more than 1 meter of uplift. Small to moderate earthquakes—more than 90,000—and another half-meter of uplift were recorded between 1983 and 1984 (Figure 161). Since the geologists knew the violent history of the caldera, this unrest prompted disaster plans and practice evacuations and also attracted volcanologists from all over the world. Then the earthquakes dropped from 10,000 back to tens or hundreds per month, and the uplift stopped. Fortunately, though, the RVO kept up its monitoring, practice evacuations, and disaster contingency plans.

Uplift began again in 1992, although without a dramatic increase in earthquakes. Then, on September 18, 1994, two strongly felt quakes occurred nearly simultaneously beneath the caldera, producing a small tsunami in the harbor. A large uplift occurred during the night of September 18–19 but was not detected until morning. By that time, Tavurvur Volcano was already erupting. With only twenty-seven hours of seismic warning, the long-expected eruption had begun. Just over an hour after the start of the Tavurvur eruption, Vulcan joined in from across the bay (Figure 162).

The combined eruptions dumped volcanic ash on Rabaul Town in layers 10 centimeters to more than 1 meter thick. Pyroclastic flows from Vulcan swept into the bay, generating tsunamis that reached as far as 200 meters onshore around parts of the harbor. Vulcan's eruption

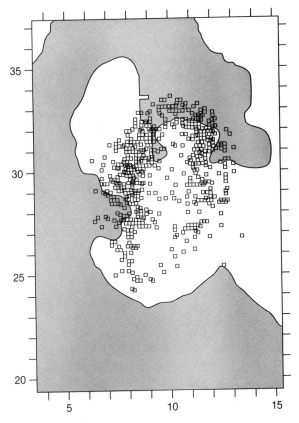

161 The locations of shallow earthquakes, most of them at depths of 3 kilometers or less, beneath Rabaul Caldera in Papua New Guinea, from September 1983 to November 1984. The area of maximum uplift was within the oval of earthquakes. The numbers along the axes indicate distance in kilometers. (From C. O. McKee, *Volcano News* 19–20 [1985] 3.)

was over in two weeks, but at Tavurvur small, intermittent eruptions have continued for many months.

Based on the earthquakes, a Stage 2 Alert (hazard situation possible) had been declared at 7:20 P.M. on September 18, eleven hours before the eruption began. A Stage 3 Alert (hazard situation probable) was declared only minutes before the eruption began. Stage 4 (evacuation required) was never announced, largely because most residents of Rabaul had begun to evacuate voluntarily during the night and early

162   View of the Rabaul, Papua New Guinea eruption from Space Shuttle Discovery on September 19, 1994. The high eruption cloud from Vulcan Volcano reaches an altitude of 18 kilometers. More than 50,000 people were evacuated from the area. (Photograph courtesy of NASA.)

morning before the eruption started. Prompted by numerous sharp earthquakes, the majority of residents had fled to safer areas before 5:30 A.M. By 9:00 A.M., three hours after the eruption had begun, Rabaul was empty.

Destruction from ashfall was extensive in the city, with its south and east sides listed as 100 percent damaged. Where the ashfall was more than half a meter thick, almost all buildings collapsed, but where it was less than 20 centimeters thick, most roofs survived. Heavy rainfall on the loose ash caused lahars, and tsunamis severely damaged the wharf area. Pyroclastic flows devastated the area around Vulcan but did not reach Rabaul. Luckily, the 750 people who lived near Vulcan had evacuated early.

Property damage from the 1994 Rabaul eruption totaled more than $200 million, but only eight people died, three of them in car accidents. With the short warning time and no mandatory evacuation, it was remarkable that the casualties were so few. Two reasons seem to have been responsible. First, many older and respected people remembered the deadly 1937 eruption. Second, the 1984 earthquake crisis had not been forgotten, even though that time it stopped without an eruption. People remembered the disaster plans and practice evacuations and fled without being ordered to do so.

After the immediate crisis of a volcanic eruption has passed, serious problems usually remain. Large numbers of people who have had to evacuate sometimes remain displaced for years, because either their homes have been destroyed, or continued volcanic activity threatens the evacuated area. Often large areas of croplands are destroyed, as happened at the Tambora eruption in Indonesia in 1815, where starvation killed an estimated 80,000 people.

Since the Pinatubo eruption in 1991, the worst continuing threat has been from the lahars that pour down the mountain each rainy season, carrying prodigious amounts of ash and pumice that was deposited in that eruption. According to a 1993 report by Philippine and U.S. Geological Survey volcanologists, "thick lahar deposits have left 50,000 people homeless, and flooding and isolation have affected more than 1,350,000 in 39 towns and four large cities. More than 1,000 square kilometers of prime agricultural land is affected by or at risk from lahars, flooding or siltation."

Recent support for reducing volcanic risk has come from an unexpected source. Near-fatal aircraft accidents resulted when jetliners temporarily lost power after flying into ash clouds from Galunggung Volcano in Indonesia in 1982, Redoubt Volcano in Alaska in 1989, and Pinatubo Volcano in 1991. Since then, scientists and the international airline industry have joined forces to identify hazardous volcanic clouds as they form and to detour aircraft around them.

A recent experiment by William Rose and David Schneider of Michigan Technological University indicates that volcanic ash clouds can be monitored from space. Using weather satellite observations of an ash eruption from Popocatépetl Volcano in Mexico in March 1996, they showed that by using images of different frequency bands from the satellite, they could discriminate between the ash cloud and weather clouds.

What will really reduce volcanic risk over the next century is not so much new concepts and techniques—although these will surely come about—but applying currently known methods. Less than 10 percent of the world's 1300 potentially active volcanoes have been mapped to appraise their hazards, and even fewer are being monitored continuously. Educating the people at risk is also necessary; many people who live in Seattle will tell you that Mount Rainier is an extinct volcano. As this book is being written, we have the ways to reduce volcanic risk but not the means.

But reducing volcanic risk is only part of understanding how volcanoes work. Volcanology is a fledgling science. It has come a long way, but it has an even longer way to go on the path to a true understanding of how volcanoes work. This book is dedicated to those who will seek that path.

Sources

Blong, Russell, and C. McKee. *The Rabaul Eruption 1994.* Sydney: Natural Hazards Research Centre, Macquarie University, 1995.

Casadevall, Tom. "Volcanic Ash and Aviation Safety." *U.S. Geological Survey Bulletin* 2047 (1994).

Newhall, C. G., and R. S. Punongbayan. eds. *Fire and Mud: Eruptions and Lahars of Mount Pinatubo, Philippines.* Seattle: University of Washington Press, 1997.

Peterson, Donald, and R. I. Tilling. "Interactions Between Scientists, Civil Authorities and the Public at Hazardous Volcanoes." In C. R. J. Kilburn and G. Luongo, eds., *Active Lavas*. London: UCL Press, 1993.

Punongbayan, R. S., et al. "Lessons from a Major Eruption: Mt. Pinatubo, Philippines." *EOS* (American Geophysical Union), December 3, 1991.

Rose, W. I., and D. J. Schneider. "Satellite Images Offer Aircraft Protection from Volcanic Ash Clouds." *EOS* (American Geophysical Union), December 10, 1996.

Scarpa, R., and R. I. Tilling, eds. *Monitoring and Mitigation of Volcano Hazards*. Berlin: Springer-Verlag, 1996.

Tilling, Robert. "Volcanic Hazards and Their Mitigation: Progress and Problems." *Reviews of Geophysics* 27 (1989).

Tilling, Robert, and P. W. Lipman. "Lessons in Reducing Volcanic Risk." *Nature*, July 22, 1993.

Williams, Stanley. "Double Trouble." *Earth*, August 1996.

APPENDIX A

The World's 101 Most Notorious Volcanoes

Activity, size, shape, beauty, danger, location, and the authors' preferences were used to select the following list. More complete catalogs of active volcanoes from which some of the data in this appendix are derived can be found in the Top Five web sites listed in Appendix B. The eruption data are through 1996.

ANTARCTICA

Mount Erebus   A snow- and ice-covered stratovolcano with an active lava lake in its summit crater. It was in eruption when first sighted in 1841 and has had several reported explosions since. The present lava lake, about 100 meters in diameter, apparently formed in the 1960s. Erebus is the world's farthest-south active volcano.

ATLANTIC OCEAN

Beerenberg   A stratovolcano with a caldera, on Jan Mayen Island north of Iceland. It is the farthest-north active volcano and has erupted five times since 1633, the last time in 1985.

Fogo   A stratovolcano with an 8-kilometer-wide caldera, in the Cape Verde Islands. It was continuously active from the time of settlement about 1500 to 1760. Ten eruptions have occurred since then, the last in 1995. That eruption covered 6 square kilometers, with about 80 million cubic meters of lava, destroying valuable farmland.

La Palma   A complex volcano with a caldera and rift zones, on the west end of the Canary Islands. Lava flows have erupted six times since 1585, the last time in 1971.

CARIBBEAN SEA

Mont Pelée   A stratovolcano with summit domes, on Martinique. The explosive eruption on May 8, 1902, generated a nuée ardente that swept down the mountainside and, within minutes, incinerated the town of Saint Pierre and its 28,000 inhabitants. Three other explosive eruptions have been recorded, the last in 1932 (see Figure 81).

La Soufrière of Guadeloupe   A stratovolcano with a summit dome, on the southern part of Guadeloupe Island. It has erupted explosively about ten times since 1400. The last eruption in 1976 prompted the evacuation of 70,000 people for several months, but the eruption turned out to be only minor explosions.

La Soufrière of St. Vincent   A stratovolcano with a crater lake. Its seven eruptions since 1718 include two major explosive events. Evacuation in 1979 prevented a potential repeat of the tragedy of 1902, when 1600 deaths occurred, largely from nuées ardentes.

Soufrière Hills of Montserrat   A stratovolcano with lava domes on the island of Montserrat. The first recorded eruption of this volcano began in July 1995. The growing lava dome has collapsed and exploded several times, producing pyroclastic flows. More than 6000 people have been evacuated to the north end of the island. This is potentially a very dangerous eruption, and the remaining residents of the island remain on high alert.

CHILE

Cerro Hudson   A stratovolcano in southern Chile. Known eruptions occurred in 1891 and 1971, and a major explosive eruption in 1991. The region near the volcano is sparsely populated, so no people were killed by the last eruption, but thousands of sheep and cattle died from fluorine poisoning from the ash-covered grasslands.

Llullaillaco   A stratovolcano in northern Chile. This is the world's highest active volcano, with a summit elevation of 6739 meters. Of the three recorded eruptions during the 1800s, two were explosive, and the other was a lava flow from a side vent. Nevado Ojos del Salado (6886 meters), also in Chile, is higher but has only steam vents and no recorded eruptions.

Villarrica   A stratovolcano in central Chile. About twenty-one explosive eruptions and three lava flows have occurred since 1558. The eruption in 1971 melted large volumes of snow and ice, and the resulting mudflows killed fifteen people. The last eruption was in 1995.

COLOMBIA

Galeras   A complex volcano in southern Colombia. It has erupted more than twenty times since 1535. The last explosive eruption in 1993 was small but killed nine people, including six volcanologists who were in the crater. It is clear that forecasting volcanic eruptions still has a long way to go.

Nevado del Ruiz   A high (5321 meters) stratovolcano in central Colombia. Mudflows from an earthquake or small eruption in 1845 killed about 1000 people. Another small-to-moderate eruption in 1985 melted part of the summit ice cap; the resulting mudflows caused about 23,000 fatalities (see Chapter 10).

COSTA RICA

Arenal   A stratovolcano in northwestern Costa Rica. Dormant until 1968, it burst into eruption with a strong explosion throwing huge blocks as far as 5 kilometers; the explosion was followed by nuées ardentes. This initial eruption killed seventy-eight people. Activity has continued since, with the extrusion of thick, slow-moving lava flows.

Irazú   A stratovolcano with a double crater, in central Costa Rica. It has erupted explosively about fifteen times since 1723. The last major eruption lasted from 1963 until 1965; its numerous small-to-moderate ashfalls were destructive to coffee plantations and a nuisance to the capital city of San Juan.

Poás   A stratovolcano with twin crater lakes, in central Costa Rica. It has erupted twenty-four times since 1834, generally with explosions of mud and water from the northern lake. The 1910 eruption shot a fountain of water more than 4 kilometers high. The last eruption was in 1994.

ECUADOR (ANDES MOUNTAINS)

Cotopaxi   A stratovolcano nearly 6000 meters high. It has had more than fifty eruptions since 1532, including major eruptions and lava flows. The 1877 eruption melted large volumes of snow and ice from the summit, causing mudflows that traveled 100 kilometers down adjacent riverbeds. The last eruption was in 1942 (see Figure 95).

Guagua Pichincha   A stratovolcano with a caldera and a central cone. Although active from the 1500s to the 1800s, there have been only small eruptions in this century, the last one in 1993. Forty centimeters of ash fell on the capital city of Quito in 1660; modern Quito climbs up the sides of this potentially dangerous volcano.

Reventador   A stratovolcano east of the main ranges of the Andes. Although this mysterious volcano was not explored until 1931, it was apparently the source of some ashfalls in Quito as far back as 1541. An important oil pipeline connecting the Amazon fields to the west coast of Ecuador now crosses its north slope. The last eruption was in 1976.

Sangay   A 5230-meter-high snow- and ice-covered stratovolcano. Small explosive eruptions with occasional lava flows kept this volcano in nearly continuous activity from 1728 to 1916 and from 1934 to 1996.

ECUADOR (GALAPAGOS ISLANDS)

Fernandina   A shield volcano with a summit caldera. The most active of the Galapagos volcanoes, this uninhabited island was called Narborough in Darwin's chronicles. A major explosive eruption and 350-meter collapse of the caldera occurred in 1968. The last eruption was in 1995.

EL SALVADOR

Izalco   A young stratovolcano born in 1770 on the south flank of Santa Ana Volcano in western El Salvador. It had nearly continuous small explosive eruptions until 1957 and was known as the Lighthouse of the Pacific.

When a hotel was built nearby to view the frequent eruptions, the activity stopped. The last eruption was in 1966.

ETHIOPIA

Erta Alè   A shield volcano with an active lava lake, in the rift valley of northern Ethiopia. Lava eruptions from fissures on the flank were observed in 1959–1960. The active lava lake discovered in 1967 has since been in constant eruption.

GREECE

Santorini   A stratovolcano with a submerged caldera, in the Aegean Sea. Its giant explosive eruption and caldera collapse in about 1600 B.C. buried Akroteri, an important Minoan city currently under excavation. The huge eruption and sudden sinking of the island's center beneath the sea may have been the source of the legend of Atlantis. The last eruption in 1950 formed a lava dome and thick lava flows on the islands within the caldera.

GUATEMALA

Atitlán   A stratovolcano on the south rim of a 20-kilometer-diameter caldera lake, in southwestern Guatemala. It has erupted with small to moderate explosions eleven times since 1469. The caldera lake, formed following immense prehistoric eruptions, is one of the most beautiful in the world.

Fuego   A stratovolcano—whose name means fire—in southwestern Guatemala. It has erupted more than fifty times since 1524, producing mostly explosions of ash but sometimes nuées ardentes and lava flows. The last eruption was in 1987.

Pacaya   A volcanic complex of two small stratovolcano cones and older lava domes, in southern Guatemala. It has erupted more than thirty-five times since 1565, generally only with explosions but with some lava flows in recent years. Eruptions have occurred nearly every year since 1965.

Santa Maria   A stratovolcano with a growing lava dome on its southwest slope, in western Guatemala. Its first eruption in historic time was in

1902; it produced a giant explosion of 5.5 cubic kilometers of pumice fragments and ash. A lava dome named Santiaguito began growing in the explosion crater in 1922 and has since been erupting intermittently. The last eruption was in 1996.

ICELAND

Askja   A complex volcano with a 10-kilometer-diameter caldera, in central Iceland. Although most of its eight eruptions since the fourteenth century have been lava flows, the great explosion of 1875 showered 2 cubic kilometers of ash over much of eastern Iceland. The resulting famine led many Icelanders to emigrate to the United States and Canada. The last eruption was in 1961.

Grímsvötn   A caldera in the Vatnajökull ice cap of south central Iceland. It has erupted beneath the ice about twenty-five times since 1332, causing gigantic floods called *glacial bursts*. The sudden floods often exceed the flow volume of the Mississippi River. The last eruption was in 1996.

Heimaey   A cinder cone with a thick, blocky, lava flow from its north side, in the Vestmann Islands off the south coast of Iceland. A 2-kilometer-long fissure opened near the fishing port of Heimaey in 1973. Cinders and ash soon covered much of the evacuated town of 5000 inhabitants, and the thick flow nearly closed off the harbor entrance. The courageous Icelanders returned to a better harbor, rebuilt their town, and even heated their hospital with steam from the cooling flows.

Hekla   A stratovolcano elongated by a northeast-trending rift system, in south central Iceland. It has erupted twenty times since the settling of Iceland in A.D. 900, generally with ash explosions followed by lava flows. In medieval Europe, Hekla was considered the gate to hell. The last eruption was in 1991.

Krafla   A complex volcano with a large central caldera, in northern Iceland. Dormant after a series of eruptions between 1724 and 1728, Krafla awakened in 1975 with small eruptions and episodes of extensive ground cracking. An eruption in 1977 also sprayed a small amount of lava out of a producing geothermal steam well, the only known case of an eruption from a man-made vent (see Figure 30). The last eruption was in 1984.

Laki   A fissure zone more than 25 kilometers long, in south central Iceland. Its single giant eruption in 1783 produced more than 12 cubic kilometers of lava, a historic record, filling two river valleys and covering more than 500 square kilometers. Stunted grass and fluorine poisoning from the accompanying volcanic gases starved and killed most of Iceland's livestock. The ensuing famine caused 10,000 deaths (see Figures 86 and 139).

Surtsey   A cinder cone and lava flow island on the south coast of Iceland. Born from the sea in 1963 and erupting until 1967, it has provided scientists a view of how land forms and plants and animals establish themselves in this new territory (see Chapter 2).

INDIAN OCEAN

Karthala   A shield volcano with a summit caldera, on the southern part of Grand Comoro Island. It has erupted about twenty times since 1828, mainly lava flows issuing from the summit or rift zones. The last eruption was in 1991.

Piton de la Fournaise   A shield volcano with a caldera, on the eastern part of Reunion Island. Sometimes called a sister to Hawaiian volcanoes because of the similarity of climate and volcanic nature, it has erupted lava flows more than 100 times since 1640. The last eruption was in 1992.

INDONESIA

Agung   A stratovolcano in Bali considered in legend to be the navel of the world. Although it has erupted explosively only four times since 1808, the last eruption in 1963–1964 was of major proportions. High ash explosions affected world climate, and pyroclastic flows killed 2000 people.

Colo   A low (508 meters) stratovolcano with a 2-kilometer-wide caldera and small crater lake, on Una Una Island near Sulawesi (Celebes). A large explosive eruption occurred in 1983 just after the 7000 inhabitants of the island had been evacuated (see Chapter 18).

Dieng   A complex volcanic plateau with twenty-six cones and craters, in central Java. It has erupted about ten times since 1825; the eruption in 1979 consisted of poisonous gases that killed 150 people. The last eruption was in 1993.

Galunggung   A stratovolcano with a lava dome, in western Java. It has erupted only five times in history, but the first in 1822 produced a

22-kilometer-long debris avalanche and mudflow that killed 4000 people. The 1982 eruption killed 27 people. The last eruption was in 1984.

Gamalama   A stratovolcano island with multiple craters and crater lakes, west of Halmahera. It has erupted explosively more than sixty times since 1538, sometimes producing lava flows. The last eruption was in 1994.

Kelut   A stratovolcano with a crater lake, in eastern Java. It has erupted thirty times since about A.D. 1000. The explosive eruptions throughout the hot crater lake and cause widespread destruction. In the 1919 eruption, more than 100 villages were destroyed or damaged by mudflows, which killed 5100 people. The last eruption was in 1990.

Krakatau   Stratovolcano islands around a submerged caldera, in the Sunda Strait between Sumatra and Java. The 1883 eruption was one of the largest natural explosions in recorded time. Sounds were heard for 4000 kilometers; the emitted ash and pumice blocks totaled 18 cubic kilometers; the 6-kilometer-diameter caldera collapsed; and the resulting tsunamis killed 36,000 people on the low shores of Java and Sumatra. The last eruption was in 1996 (see Chapter 4).

Merapi   A stratovolcano—whose name means mountain of fire—with a summit lava dome, in central Java. It has erupted more than sixty times since A.D. 1006, generally with explosions and nuées ardentes. The 1006 eruption caused so much death and destruction that the Hindu rajah moved to Bali, and Java became Muslim. The last eruption was in 1996.

Semeru   A stratovolcano with a summit lava dome, in eastern Java. Its nearly seventy eruptions—generally explosions sometimes accompanied by pyroclastic flows and lava flows—make it one of Java's most active as well as its highest (3620 meters) volcano. The last eruption was in 1996.

Tambora   A stratovolcano with a summit caldera, on Sumbawa Island. Its giant eruption in 1815 exceeded the size and power of Krakatau. The explosion, followed by caldera collapse, is estimated to have produced about 40 cubic kilometers of ash and blocks. Ten thousand people were killed by the eruption, and 80,000 starved in the resulting crop loss and famine. World climate may have been affected (see Chapter 16). The last eruption was in 1967.

ITALY

Etna   A transitional shield-to-stratovolcano in northeastern Sicily. It has erupted lava flows more than 150 times since activity was first recorded in 1500 B.C. Small-to-moderate explosive eruptions occur at the summit; one in 1979 took nine lives. The last eruption was in 1996.

Stromboli   A stratovolcano island west of Italy. Known as the Lighthouse of the Mediterranean, it has been in almost continuous eruption for more than 2000 years. Small explosions of incandescent lava hurled up from the crater every 15 to 30 minutes are visible to ships passing by. Larger eruptions, some with lava flows, take place every several years.

Vesuvius   A complex stratovolcano east of Naples. Most famous for its A.D. 79 eruption that buried Pompeii, it has since erupted more than fifty times. Explosive eruptions are generally followed by lava flows. The last eruption was in 1944 (see Figure 138).

Vulcano   A stratovolcano west of Italy. Legendary forge of Vulcan, this small island has provided the family name for all volcanoes. It has erupted explosively about ten times since 200 B.C., the last time in 1890.

JAPAN

Asama   A complex stratovolcano in central Japan. It has erupted more than 100 times since A.D. 685, generally with a series of small explosions. In 1783 large pyroclastic flows and mudflows buried villages, killing 1300 people. Scientists at the Asama Volcano Observatory, established in 1909, have found earthquake counts and locations to be useful in forecasting eruptions. The last eruption was in 1990.

Aso   A group of cinder cones and small stratovolcanoes within a 20-kilometer-diameter caldera. One vent has erupted more than 100 times since A.D. 796, generally as single, isolated explosions. Tourists visiting the rim of the active vent are sometimes killed by ejected blocks and bombs. The last eruption was in 1996.

Bandai   A stratovolcano with a caldera, in north central Japan. It has erupted six times since A.D. 806, including the major explosions that formed the caldera in 1888. The debris from this avalanche and explosion, amounting to 1.2 cubic kilometers, swept down the north flank of Bandai, burying villages and killing 460 people.

Bayonnaise Rocks   Lava domes that sometimes build ephemeral islands inside a submarine caldera southeast of Japan. This submarine volcanic center has erupted more than ten times since it was first witnessed in 1896, sometimes building an island up to 95 meters high. A Japanese oceanographic ship investigating the area was destroyed by an explosion in 1952, with the loss of all thirty-one persons on board. The last eruption was in 1988.

Fuji   A classic stratovolcano in central Japan, the archetype of volcanic form. It has erupted ash and lava about fifteen times since A.D. 781. The last eruption in 1707 was from a vent high on the southeast side that ejected 0.8 cubic kilometer of ash, blocks, and bombs. The finer ash reached Tokyo (see Figure 54).

Oshima   A stratovolcano island with a caldera and a central cone, off the east coast of central Japan. It has erupted about fifty times since A.D. 684, sometimes explosively and sometimes with extensive lava flows. The last eruption was in 1990.

Sakurajima   A stratovolcano forming a peninsula into Kagoshima Bay, in southern Japan. One of the world's most active volcanoes, it has had thousands of small explosive eruptions since the first recorded one in A.D. 708. Many deaths occurred in 1476 and 1779, but sufficient warning from earthquakes kept the death toll to only a few persons in the giant eruption of 1914, which produced 0.6 cubic kilometer of ash and 1.6 cubic kilometers of thick lava flows. The last eruption was in 1996.

Unzen   A complex volcanic peninsula formed by several lava domes, on the west coast of southern Japan. Although only six eruptions have been recorded since A.D. 860, the 1792 activity involved either an explosion or earthquake that triggered a 0.5-cubic-kilometer avalanche. The slide and resulting tsunami caused 14,000 deaths. The 1991 eruption killed forty-three people, including three volcanologists (see Figure 157).

Usu   A stratovolcano with summit and flank lava domes, in northern Japan. Although it has erupted only seven times since 1663, most of these eruptions have been explosive and destructive to life or land. In 1910, 1943–1944, and 1977–1982, large lava domes or ground surface uplifts—kilometers in diameter and tens to hundreds of meters high—were slowly forced upward.

MEXICO

El Chichón   A small stratovolcano in Chiapas State in southern Mexico, dormant until 1982. Its sudden explosive eruption generated pyroclastic flows that killed about 2500 people and injected a huge cloud of dust and sulfuric acid aerosol into the stratosphere (see Chapter 16).

Colima   A stratovolcano with a summit lava dome, in west central Mexico. Its more than forty eruptions since 1576 have been mostly explosive, with some lava flows. The last eruption was in 1994.

Parícutin   A cinder cone in central Mexico. Born in a cornfield in 1943, it built a 410-meter-high cone with extensive lava fields during its brief life span. Most of the 1.3 cubic kilometers of ash and cinders and much of the 0.7 cubic kilometer of lava were produced in the first few years. The eruption ended in 1952.

Popocatépetl   A snow-capped stratovolcano dominating the skyline south of Mexico City. It erupted with small explosions eleven times between 1512 and 1697. Since a major explosion in 1720, only four small eruptions have occurred, the last in 1996.

NEW ZEALAND

Ngauruhoe   A stratovolcano with a nearly perfect cone, in central North Island. It has had more than fifty explosive eruptions since 1839, some with small pyroclastic flows. The latest eruption was in 1977 (see Figure 107).

Ruapehu   A stratovolcano with a hot crater lake, in central North Island. There have been more than thirty small explosions of steam and ash from the crater lake since 1861. Mudflows from the erupting lake occasionally flood adjacent valleys. One such flow on Christmas Eve in 1953 swept away a railroad bridge, wrecking the Wellington-Auckland Express and causing 151 deaths. The latest eruption was in 1996 (see Figure 107).

Tarawera   A volcanic complex of rhyolite domes, in central North Island. A great eruption along a 17-kilometer-long fissure in 1886 ejected 1.3 cubic kilometers of basaltic ash and hot mud, burying three villages and killing more than 150 persons. From 1900 to 1904, Waimangu (black water)

Geyser erupted in one of the craters formed along the 1886 fissure. Occasional geyser bursts of 400 to 500 meters during that time were the highest ever recorded.

White Island   A stratovolcano in the Bay of Plenty north of New Zealand, with a large horseshoe-shaped crater. About thirty-five small-to-moderate explosive eruptions have occurred since 1826. Eleven men were killed, and a sulfur works was destroyed in the 1914 eruption and landslide. The last eruption was in 1995.

NICARAGUA

Cerro Negro   A cinder cone—whose name means black hill—in western Nicaragua. Born in 1850, it has erupted about twenty times in its brief life span. Explosive eruptions from the central crater are often accompanied by lava flows issuing from near the base of the cinder cone. The last eruption was in 1995 (see Figure 49).

Coseguina   A stratovolcano with a central caldera lake, in western Nicaragua. A single major explosive eruption occurred in 1835 that scattered ash throughout Central America and southern Mexico. The ash cloud blotted out the sun over an area 300 kilometers wide. The last eruption was in 1859.

Masaya   A caldera with small central stratovolcanoes, in southwestern Nicaragua. It has erupted many times since 1524 with varied activity, including explosions, lava flows, and lava lakes. The last eruption was in 1996 (see Figure 101).

PACIFIC OCEAN

Ambrym   A stratovolcano within a huge caldera in the New Hebrides Islands. Although the record is incomplete, it has been very active since its discovery in 1774, with many ash explosions and lava flows. The last eruption was in 1996.

Macdonald Seamount   A submarine volcano at the southeast end of the Austral Islands and Seamounts in the South Pacific. Its summit is about 30 meters below sea level. This volcano was discovered to be active in 1967, when noises from it were picked up by distant hydrophones. In 1987 an oceanographic expedition sailing over the seamount witnessed the eruption of large gas bubbles and small fragments of basalt. Seismometers and hydrophones indicated that the eruptions continued into 1989.

Pagan   A stratovolcano in the Mariana Islands of the western Pacific Ocean. After four small eruptions between 1909 and 1925, a large explosive eruption in 1981 lofted an ash cloud to 20 kilometers above the island. Smaller eruptions have occurred since, the last in 1993.

Raoul   A complex stratovolcano island with a caldera and crater lakes, in the Kermadec Islands north of New Zealand. This remote volcano is noteworthy because a research seismic station there, part of a worldwide net, gave warning of the 1964 eruption by recording a forerunning swarm of local earthquakes.

Yasour   A stratovolcano in the New Hebrides Islands. It has been in almost constant mild eruption since its discovery in 1774. Numerous small explosions hurl lava bombs 20 to 200 meters into the air.

PAPUA NEW GUINEA

Karkar   A stratovolcano island with a summit caldera, off the north coast of Papua New Guinea. It has erupted about ten times since 1643. The eruption in 1979 took the lives of volcanologists Robin Cooke and Elias Ravian. The last eruption was in 1980.

Lamington   A stratovolcano with a lava dome, in eastern Papua New Guinea. This dormant volcano suddenly exploded in 1951. The pyroclastic flows devastated more than 200 square kilometers and killed about 3000 people. A 500-meter-high lava dome grew in the explosion crater from 1951 to 1956. The volcano has been quiet since then.

Manam   A 1300-meter-high stratovolcano that forms a circular island 10 kilometers in diameter, off the north coast of Papua New Guinea. intermittent small explosive eruptions occurred frequently from 1974 to 1996.

Rabaul   A group of small volcanoes around the rim of a caldera bay on New Britain Island. Although the vents around Rabaul Harbor have erupted only seven times since 1767, the violent explosive eruption in 1937 formed Vulcan Crater, generated tsunamis, and killed 500 people. The last eruption in 1995 killed 8 people; more than 50,000 had evacuated (see Figure 160).

PERU

El Misti   A classic stratovolcano rising high above the city of Arequipa. Small explosive eruptions, the last in 1870, and gentle steaming characterize its historic activity. It is noted for its beauty rather than its habits.

PHILIPPINES

Mayon   A classic stratovolcano cone in the central Philippines. It has erupted explosively more than forty times since 1616, often producing pyroclastic flows and lava flows. The region is densely populated, and at least nine eruptions have caused fatalities. The last eruption was in 1995 (see Figure 100).

Pinatubo   An eroded stratovolcano on Luzon Island in the northern Philippines. Until 1991 this volcano had no recorded eruptions; this was suddenly changed by the second largest eruption of this century (after Katmai). Huge pyroclastic flows and mudflows inundated the surrounding areas. More than 300 people were killed by the eruption, but tens of thousands more would have been killed if the region had not been evacuated. This giant eruption is a classic success story of eruption forecasting (see Figure 158 and Color Plate 17).

Taal   A stratovolcano island within a huge caldera lake, in the central Philippines. It has erupted more than thirty times since 1572, generally explosions sometimes accompanied by tsunamis in the lake. This killer volcano took many lives in 1716, 1749, 1754, 1874, 1911, and 1965. The last eruption was in 1977.

RUSSIA (KAMCHATKA)

Bezymianny   A complex stratovolcano in central Kamchatka. Small explosive eruptions beginning in 1955 climaxed in a gigantic explosion in 1956, forming a mushroom cloud 35 kilometers high (see Figure 56). The thick, hot ash flow deposits formed the Valley of Ten Thousand Smokes of Kamchatka, a feature similar to that created by the Katmai eruption in Alaska in 1912. The last eruption was in 1996.

Karymsky   A stratovolcano within an ancient caldera, in southeast Kamchatka. Its thirty-six eruptions since 1771 were mostly explosive. The last eruption was in 1996.

Kliuchevskoi   A classic stratovolcano in central Kamchatka. It has erupted explosively more than eighty times since 1697, often producing lava flows. The last eruption was in 1996.

Tolbachik   A stratovolcano with a caldera, in central Kamchatka. After about twenty moderate eruptions since 1740, it produced a giant fissure eruption in 1975 that built several large cinder cones and major lava flows totaling nearly 2 cubic kilometers in volume. Forerunning earthquake swarms led to a prediction of the time and place of the 1975 eruption, precise enough that TV camera crews were on location for the outbreak (see Figure 155).

TANZANIA

Ol Doinyo Lengai   A stratovolcano—whose name means mountain of God—in northern Tanzania. Although it has erupted more than ten times since 1880, this volcano is more famous to science for its unusual lava flows composed partly of sodium carbonate. The last eruption was in 1996 (see Figure 137).

UNITED STATES (ALASKA)

Augustine   A stratovolcano island with a summit lava dome, in the Cook Inlet, southern Alaska. It has erupted explosively seven times since 1812. Dangers from major ashfalls or volcanically generated tsunamis worry the offshore oil developers located nearby. The last eruption was in 1986.

Katmai   A complex stratovolcano with a summit caldera and lake, on the Alaska Peninsula. The 1912 eruption of Katmai was among the world's largest in historic time. In two days, thick ash deposits covered a huge area, and a pyroclastic flow filled a valley 3 kilometers wide and 20 kilometers long creating the Valley of Ten Thousand Smokes. The summit caldera, 5 kilometers in diameter, was formed by collapse during the eruption. The total volume of the ash and glowing avalanche was 30 cubic kilometers. Although most of the valley "smokes" (steam vents) are now dead, Mount Trident, a new vent on the west flank of Katmai, became active in 1949 and has exploded and extruded thick lava flows several times. The last eruption of Trident was in 1974.

Pavlof   A stratovolcano on the Alaska Peninsula (see Figure 94). It has had more than forty periods of small-to-moderate explosive eruptions since 1790. The last eruption was in 1996.

Redoubt   A stratovolcano in southwestern Alaska. This peak, visible from the city of Anchorage, has erupted only a few times in recorded history. The 1989–1990 explosive eruptions generated mudflows that threatened oil storage facilities on the north shore of Cook Inlet. The ash cloud almost caused the crash of a large airliner by clogging its jet engines.

Shishaldin   A classic stratovolcano cone in the eastern Aleutian Islands. It has a record of about thirty small-to-moderate explosive eruptions since 1775. Its sharp, snow-covered peak often has a plume of volcanic gases. The last eruption was in 1995.

UNITED STATES (CASCADE RANGE)

Lassen Peak   A stratovolcano with a summit lava dome, in northern California. It had a series of explosive eruptions in 1914–1915 that culminated in avalanches, a pyroclastic flow, and mudflows in May 1915. Small explosions continued until 1917 (see Figure 97).

Mount Rainier   A massive, glacier-covered stratovolcano near Seattle, Washington. It has been dormant since about 1850, when a small explosive eruption apparently occurred. Small steam vents issue from the edges of the snow- and ice-filled summit crater (see Figures 91 and Color Plate 19)

Mount St. Helens   A snow-covered stratovolcano in southern Washington known for its beauty and serenity before its gigantic explosive eruption on May 18, 1980. A directed blast leveled 500 square kilometers of forest, and a major debris avalanche filled a valley for 25 kilometers. The 2950-meter-high summit was lowered by 400 meters, forming a deep horseshoe crater facing north. About fifty-seven people were killed, including David Johnston, a volcanologist. The last eruption, a steam explosion, was in 1991 (see Chapter 5).

UNITED STATES (HAWAII)

Kilauea   A shield volcano with a summit caldera, on the island of Hawaii. Famous for its active lava lake during the 1800s and early 1900s, it has also erupted extensive lava flows more than fifty times from both the summit and rift zones. The last eruption was in 1996 (see Chapter 7 for additional information on Kilauea's Puu Oo eruption, which began in 1983).

Mauna Loa   A massive shield volcano with a summit caldera, on the island of Hawaii. It has erupted large lava flows from both the summit and rift zones thirty-eight times since 1832, producing a total of nearly 4 cubic kilometers of basalt. The last eruption was in 1984 (see Figures 106, 145, 153, and 154).

ZAIRE

Nyamuragira   A shield volcano with a summit caldera, in eastern Zaire. It has erupted more than twenty times since 1894, including lava lake activity from 1921 to 1938. The last eruption was in 1996.

Nyiragongo   A stratovolcano with a summit caldera, in eastern Zaire. It has erupted about fifteen times since 1884, including lava lake activity from 1935 to 1977. A major fissure eruption on the south flank in 1977 drained the lava lake and rapidly covered an area of several square kilometers with very fluid lavas. About 300 people were killed by these flows. The last eruption was in 1995 (see Color Plate 23).

APPENDIX B

Volcanoes on the Internet

Many institutions and individuals have created excellent home pages that provide information about volcanoes, new eruption data, great photographs, video clips, and even profiles of volcanologists.

The following web sites are listed in geographical and topical groups. In our opinion, the Top Five are currently the best. They also provide links to many other volcano sites. These five home pages provide "master keys" that can open up much of the rest of the list without typing in the URL (Uniform Resource Locator—the address code) of the site you are trying to find.

Remember that web sites come and go, URLs change, and some sites are updated more frequently than others. Log on and explore volcanoes on the Internet. Getting there is half the fun.

TOP FIVE (AS OF 1996 IN ALPHABETICAL ORDER)

Boris Behncke's Volcano Home Page: Emphasis on Italy, enthusiastic, an overall excellent site.
   http://www.geo.mtu.edu/~boris/

Electronic Volcano (Dartmouth College):   Encyclopedic, for serious students, frequently updated, few frills.
   http://www.dartmouth.edu/~volcano/

Smithsonian Global Volcanism Program: User friendly, global view, excellent list of annotated links.
   http://www.volcano.si.edu/gvp/

Volcano World (University of North Dakota):   All ages, fine visuals, great links, frequently updated.
   http://volcano.und.nodak.edu/

Volcanoes Page (Michigan Technological University):   For serious students, excellent links, frequently updated, great visuals.
   http://www.geo.mtu.edu/volcanoes/

WORLD VOLCANOES

Earth Magazine:
   http://www.kalmbach.com/earth/earthmag.html

Geoscience Information Society:
   http://www.lib.berkeley.edu/GIS/

International Association of Volcanology (IAVCEI):
   http://geont1.lanl.gov/Heiken/ONE/IAVCEI_home_page.HTM

Los Alamos National Laboratory:
   http://bang.lanl.gov/solarsys/tervolc.htm

NASA Earth Observing System:
   http://www.geo.mtu.edu/eos/

NASA Earth Images from Space:
   http://images.jsc.nasa.gov/
   http://www.jpl.nasa.gov/archive/images.html
   http://skye.gsfc.nasa.gov/

National Geophysical Data Center (part of NOAA):
   http://www.ngdc.noaa.gov/

Natural Hazards Mitigation Group (University of Geneva):
   http://www.unige.ch/hazards/

Scott's Hotlist of Volcanoes and Earthquakes:
   http://rcch.com/hotlist/058.htm

Submarine Volcanism:
   http://www.pmel.noaa.gov/vents/

U.K. Volcanology:
   http://exodus.open.ac.uk/volcano/volc_index.html
   http://igt-www.esc.cam.ac.uk/ITG/hydrothermal/hydrothermal.html

Volcano Quarterly:
   http://members.aol.com/vqjantan

ANTARCTICA

Mount Erebus Volcano Observatory:
   http://griffy.nmt.edu/Geol/volcanology/erebus.html

LATIN AMERICAN VOLCANOES

   http://mistral.ere.umontreal.ca/~williamg/LAVolc.html

ECUADOR

Galapagos:
   http://fcdarwin.org.ec/welcome.html

ICELAND

Nordic Volcanological Institute:
   http://www.norvol.hi.is/

Science Institute, University of Iceland:
   http://www.hi.is/~mmh/gos/

ITALY

Stromboli:
   http://www.ezinfo.ethz.ch/ezinfo/volcano/strombolihomee.html

Gruppo Nazionale per la Vucanologia:
   http://www.dst.unipi.it/gnv/index.html

Istituto Internazionale di Vulcanologia Catania:
   http://www.iiv.ct.cnr.it/

JAPAN

Hakone Volcano Research Center:
   http://hakone.eri.u-tokyo.ac.jp/index.html

Volcanological Society of Japan:
   http://hakone.eri.u-tokyo.ac.ip/kazan/VSJ1E.html

NEW ZEALAND

Ruapehu:

http://www.rses.vuw.ac.nz/geophysics/geophysics.html

UNITED STATES

Alaska Volcano Observatory:

http://www.avo.alaska.edu/

Cascades Volcano Observatory:

http://vulcan.wr.usgs.gov/

Hawaii:

http://hvo.wr.usgs.gov/
http://www.soest.hawaii.edu/
http://www.satftp.soest.hawaii.edu/space/hawaii/all.bi.vfts.html
http://www.soest.hawaii.edu/GG/hcv.html
http://hiiaka.uhh.hawaii.edu/csav/start.html
http://www.bishop.hawaii.org:80/bishop/geology/
http://quake.wr.usgs.gov/QUAKES/CURRENT/hvo.html
http://www.volcanostore.com
http://www.volcanovideo.com

Long Valley Caldera:

http://quake.wr.usgs.gov/VOLCANOES/LongValley/

U.S. Geological Survey:

http://geology.usgs.gov/

Universities:

http://adder.colorado.edu/~hazctr/Home.html
ftp://129.171.100.35/pub/www/tims-home.html
http://www.vsc.washington.edu/

Yellowstone:

http://www.seis.utah.edu/HTML/YPSeismicityMaps.html
http://www.wku.edu/www/geoweb/geyser.html

INTERNET SEARCH ENGINES

If these sources don't feature the volcano or offer the information you are looking for, try the conventional search programs like AltaVista, Excite, Infoseek, Lycos, Magellan, OpenText, or Yahoo, and type in volcano, volcanoes, or the name of the specific volcano you are trying to find.

Appendix C

Metric-English Conversion Table

Length	
1 centimeter	0.3937 inch
1 inch	2.5400 centimeters
1 meter	3.2808 feet
1 foot	0.3048 meter
1 meter	1.0936 yards
1 yard	0.9144 meter
1 kilometer	0.6214 mile
1 kilometer	3281 feet
1 mile	1.6093 kilometers

Area	
1 square centimeter	0.1550 square inch
1 square inch	6.452 square centimeters
1 square meter	10.764 square feet
1 square meter	1.1960 square yards
1 square foot	0.0929 square meter
1 square kilometer	0.3861 square mile
1 square mile	2.590 square kilometers

Volume	
1 cubic centimeter	0.0610 cubic inch
1 cubic inch	16.3872 cubic centimeters
1 cubic meter	35.314 cubic feet
1 cubic foot	0.02832 cubic meter
1 cubic meter	1.3079 cubic yards
1 cubic yard	0.7646 cubic meter

Mass	
1 gram	0.03527 ounce
1 ounce	28.3495 grams
1 kilogram	2.20462 pounds
1 pound	0.45359 kilogram

Density	
1 gram/cubic centimeter	62.4280 pounds/cubic feet

Pressure	
1 kilogram/square centimeter	0.96784 atmosphere
1 kilogram/square centimeter	0.98067 bar
1 kilogram/square centimeter	14.2233 pounds/square inch
1 bar	0.98692 atmosphere
1 atmosphere	1.0332 kilogram/square cm

Temperature	
$(\text{Celsius} \times \frac{9}{5}) + 32$	Fahrenheit
$(\text{Fahrenheit} - 32) \times \frac{5}{9}$	Celsius

Energy	
1 erg	2.39006×10^{-8} gram calorie
1 joule	10^7 ergs
Explosion equivalent to 1000 tons of TNT	4×10^{19} ergs

Power	
1 watt	10^7 ergs/second
1 watt	0.001341 horsepower

Glossary

aa   A type of lava flow having a rough, fragmental surface.

active volcano   A volcano that is erupting or has erupted in recorded history.

airfall deposit   Pyroclastic fragments that have fallen from an eruption cloud.

andesite   A volcanic rock type intermediate in composition between dacite and basalt.

angle of repose   The steepest slope at which loose material will come to rest without slumping.

ash cloud   Explosive eruption cloud containing large quantities of volcanic ash.

ash flow   A pyroclastic flow of mainly ash-sized (less than 2 millimeters in diameter) material.

ash, volcanic   Fine pyroclastic material, down to dust size, that is formed by explosive volcanic eruptions.

avalanche   A large amount of snow, ice, soil, and/or rock falling or sliding at high speed.

barograph   An instrument that makes a continuous record of changes in atmospheric pressure.

basalt   A dark, heavy lava composed of about 50 percent silica, 15 percent alumina, and significant amounts of iron, calcium, and magnesium.

block, volcanic   A solid fragment thrown out in an explosive eruption, ranging in size from about 6 centimeters to several meters in diameter.

bomb, volcanic   A still-viscous lava lump thrown out in an explosive eruption, which takes on its rounded shape while in flight.

caldera   A gigantic basin with steep walls at the summit of a volcano; larger than a crater, and usually formed by collapse.

cinder cone   A steep conical hill formed by the accumulation of cinders and other loose material expelled from a volcanic vent by escaping gases.

cinder, volcanic   A pyroclastic fragment about 1 centimeter in diameter.

composite cone   See *stratovolcano*.

conduit   The pipe or crack through which magma moves.

continental crust   The solid outer layers of the Earth including granitic rocks of the continent.

continental drift   The theory that horizontal movements of the Earth's surface cause slow relative movements of continents toward or away from one another.

crater   A bowl- or funnel-shaped depression, generally in the top of a volcanic cone; often the major vent of volcanic products.

crystalline rock   A hard rock composed of interlocking crystals; often of igneous origin.

dacite   A volcanic rock type intermediate in composition between rhyolite and andesite.

debris avalanche   An avalanche composed of a mixture of rock, soil, ice, snow, trees, and so forth.

debris flow   Rapid flow of debris of various kinds, for example, a high-density mudflow. Debris avalanches become debris flows after mixing with water.

debris flow, cohesive   A debris flow with high clay content that behaves like wet concrete.

debris flow, noncohesive   A debris flow with low clay content that deposits its coarse and dense material as it loses velocity.

dike   A sheetlike body of intrusive igneous rock that cuts across the layering of the host rock.

dormant volcano   A volcano that is not presently erupting but is considered likely to do so in the future.

dust, volcanic   The finer particles of volcanic ash.

earthquake swarm   A sequence of closely spaced earthquakes during which the energy release is approximately constant (as opposed to a sequence consisting of a large quake followed by diminishing aftershocks).

earthquake wave   A general term for a vibrational wave produced by an earthquake.

effusive eruption   An eruption consisting mainly of lava flows (as opposed to explosive eruption).

eruption cloud   A gaseous cloud of volcanic ash and other pyroclastics that forms by volcanic explosion.

eruption column   The trunk or lower portion of an eruption cloud that connects the cloud to the vent.

extensional margin   The edges of tectonic plates that are moving apart.

extinct volcano   A volcano that is not erupting and is not expected to do so in the future; a dead volcano.

fault   A fracture in the Earth's crust along which there has been movement.

feldspar   A light-colored mineral composed largely of silicon, oxygen, and aluminum.

flank eruption   Eruption from the side of a volcano (in contrast to a summit eruption).

fume   A gaseous cloud without volcanic ash.

gabbro   A coarse-grained igneous rock of basaltic composition.

geophysics   The physical and mechanical aspects of geology (in contrast to geochemistry).

geothermal energy   Energy derived from the internal heat of the Earth.

geothermal power   Power generated by using the heat energy of the Earth.

granite   A coarse-grained igneous rock composed mostly of quartz and feldspar.

hazard   A potentially dangerous phenomenon like a pyroclastic flow or mudflow. See *risk*.

heat transfer   Movement of heat from one place to another.

hot-spot volcanoes   Volcanoes related to a persistent heat source in the mantle.

hydrothermal reservoir   An underground zone of porous rock containing hot water.

hyperconcentrated flow   Transitional flood between a mudflow and seasonal muddy-water stream flow.

igneous rocks   Rocks formed from the cooling and solidification of magma or lava.

ignimbrite   A rock formed by the consolidation of pyroclastic-flow deposits.

intrusion   An igneous rock body formed when molten igneous rock forces its way into surrounding host rocks and then cools; also the process of forming such an igneous rock body.

island arc   A curving chain of volcanic islands formed at a compressional plate boundary.

Kimberlite pipe   A vertical pipe-shaped intrusion of unusual igneous rocks that often contain diamonds.

lahar   An Indonesian name for a volcanic mudflow.

lapilli   Pyroclastic fragments about 1 centimeter in diameter. Lapilli is an Italian word meaning little stones.

lateral blast   A hot, low-density mixture of rock debris, ash, and gases, generated by an explosion, that moves at high speed along the ground surface.

lava   Magma or molten rock that has reached the surface; or its resulting solid rock after cooling.

lava dome   A steep-sided, rounded extrusion of highly viscous lava, squeezed out from a volcanic vent.

lava lake   A lake of molten lava in a volcanic crater or depression; refers to the solidified and partly solidified stages as well as to an active lava lake (Figure 67).

lava tube   A tunnel formed when the surface of a lava flow cools and solidifies and the still-molten interior lava flows through; also a hollow tube that remains when the interior lava drains away.

linear vent   A vent formed by a long fissure reaching the surface (in contrast to a single crater).

low-velocity layer   The zone in the upper mantle, about 60 to 250 kilometers in depth, in which seismic velocities are lower than in the overlying layers.

maar   A wide, low-relief crater formed by explosion, generally filled with water.

magma   Molten rock material with dissolved gases that forms igneous rocks on cooling; magma that reaches the surface is called lava.

magma chamber   An underground reservoir in the Earth's crust filled with magma, from which volcanic materials are derived.

magmatic fluids   Volcanic gases, especially water and carbon dioxide, dissolved in magma.

magnetic field   A region in which magnetic forces exist.

mantle   The zone of the Earth below the crust and above the core (to a depth of 3480 kilometers).

microearthquake   An earthquake that is not felt but is detectable by a seismograph.

mudflow   A flowing mixture of water-saturated mud and debris that moves downslope under the force of gravity.

neck   A vertical, pipelike intrusion that represents a former volcanic vent; usually used to describe an erosional remnant.

normal fault   An inclined fault in which the upper block moves relatively downward.

nuée ardente   A fast-moving, dense "glowing cloud" of hot volcanic ash and gas erupted from a volcano.

obsidian   A black or dark-colored volcanic glass, usually composed of rhyolite.

oceanic crust   The Earth's crust where it underlies the oceans, without the granite layer that forms continents.

olivine   An olive-green mineral composed of iron, magnesium, silicon, and oxygen.

ore   The naturally occurring material from which a mineral, or minerals, of commercial value can be extracted.

pahoehoe   A type of lava flow with a smooth, billowy, or undulating surface.

partial crystallization   The stage of cooling of magma when it is partly solid crystals and partly liquid rock.

partial melt   The stage of melting of rock when it is partly liquid rock and partly solid crystals.

pillow lava   Interconnected, sacklike bodies of lava formed under water.

plate tectonics   The theory that the Earth's crust is broken into about 12 large plates that slowly move about the surface.

plume   A rising column of magma from deep in the mantle responsible for hot-spot volcanoes; sometimes refers to ash or fume cloud.

pluton   A large igneous intrusion, formed at depth in the crust.

precipitate   A solid forming from a solution.

pumice   A form of volcanic glass so filled with gas bubble holes that it resembles a sponge and is very light.

pumice flow   A pyroclastic flow of mainly pumice fragments.

pyroclastic deposit   The material deposited by a pyroclastic flow or pyroclastic fall. The word pyroclastic should be used as an adjective.

pyroclastic fall   The fallout material from an eruption cloud, also called tephra.

pyroclastic flow   Fluidized mass of hot, dry rock fragments mixed with hot gases. Moves away from a volcano at high speeds.

pyroclastic surge   A variety of pyroclastic flow caused by base surge (a ground-hugging, expanding ring of gas and debris at the base of an explosion column).

quartz   An important rock-forming mineral composed of silicon and oxygen (SiO_2).

rhyolite   A fine-grained volcanic rock with the same composition as granite; although rhyolite and granite have the same composition, they differ in texture.

ridge, oceanic   A major submarine mountain range.

rift system   The oceanic ridges, over 60,000 kilometers in length, formed where plates are separating and new crust is being created; also their on-land equivalents like the East Africa Rift.

Ring of Fire   The regions of mountain-building earthquakes and volcanoes that surround the Pacific Ocean.

risk   The danger to life or property posed by a hazard.

seafloor spreading   The mechanism by which new seafloor crust is created at oceanic ridges and slowly spreads away on the separating plates.

seamount   An isolated tall mountain on the seafloor, generally volcanic.

sedimentary mud   Loose fine-grained sediment with enough water to form soft mud.

seismic wave   See *earthquake wave*.

seismograph   An instrument that records the motions of the Earth's surface caused by seismic waves.

seismology   The study of earthquakes, seismic waves, and the structure of the interior of the Earth.

shearing   The motion of two surfaces sliding past one another.

shield volcano   A gently sloping volcano in the shape of a flattened dome, built by flows of very fluid basaltic lava.

silica   A chemical combination of silicon and oxygen.

silicate mineral   A mineral largely composed of silicon and oxygen.

silicic   Term used to describe silica-rich volcanic rock or magma.

slump   A slow or intermittently moving landslide or submarine slide.

stock   A large igneous intrusion roughly circular in a horizontal plane.

stratovolcano   A steep volcanic cone built both by lava flows and pyroclastic eruptions.

strike-slip fault   A nearly vertical fault with side-slipping displacement.

subduction zone   The zone of convergence of two tectonic plates, one of which usually overrides the other.

tephra   A general term for all airfall pyroclastic material from a volcano; sometimes used as a synonym for all types of pyroclastic material.

thermal gradient   The rate of change of temperature with distance or depth.

thermocouple   A pair of twisted bare wires of different metals that generate a voltage when heated. A thermoelectric device used to measure molten lava temperature.

thrust fault   A gently inclined fault whose upper side moves relatively upward.

tidal wave   See *tsunami.*

transform fault   A strike-slip fault connecting the offsets of mid-ocean ridges.

tuff   A compacted pyroclastic deposit that generally contains inclusions of sedimentary material.

tuff ring   The rim of explosion debris (tuff) that surrounds a wide, shallow crater.

tsunami   A great sea wave produced by a submarine earthquake or volcanic eruption.

vein   A mineral deposit precipitated in a rock fracture.

vent   The opening at the Earth's surface through which volcanic materials issue forth.

viscosity   A measure of resistance to flow in a liquid; water has a low viscosity while honey has a high viscosity.

volcanic front   The line of volcanoes closest to the oceanic trench in island arcs like Japan.

wave-cut terrace   A level surface formed by wave erosion of coastal rocks; may appear above sea level if uplifted.

welded tuff   A hard pyroclastic rock compacted by internal heat and the pressure of overlying material.

Acknowledgments

During the early 1970s, Bob Decker, Dartmouth College, and Jim Moore, U.S. Geological Survey, wrote a book about volcanoes. It was submitted to three publishers who replied that the manuscript was too technical and too long. John Staples of W. H. Freeman and Company suggested that we rewrite the entire book in a shorter, more popular fashion. Barbara Decker, a writer of natural history and science, agreed to join the project, and Jim Moore graciously turned over the entire job to us.

During the evolution of three editions of this book we have been greatly helped by many people, and we gratefully acknowledge the following reviewers and editors:

Reviewers: Andy Barth, Indiana University; Mark Brinkley, Mississippi State University; Elwood Brooks, California State University; Allan Cox, Stanford University; John Eichelberger, University of Alaska; Albert Eggers, University of Puget Sound; Karen Grove, San Francisco State University; Otto Muller, Alfred University; William Rose, Michigan Technological University; Craig White, Boise State University; and one anonymous reviewer.

W. H. Freeman and Company personnel: Ruth Allen, Lisa Blecher, Diana Blume, John Britch, Thomas Brown, Inez Burke, Linda Chaput, Jodi Creditor, Patrick Farace, Cindy Gorman, Christine Hastings, Wendy Heredia, Holly Hodder, Bob Ishi, Melina LaMorticella, Jerry Lyons, Rebecca Mali, Susan Perepuluk, Mike Perman, Christine Rickoff, Norma Chandler Roberts, Gretchen Sauders, Erica Lane Seifert, Patrick Shriner, Allyson Siegel, Debra Siegel, John Staples, Ruth Verez, Marybeth Vlkovic, Stephen Wagley, and Judith Wilson.

Our thanks to all of you,
Robert and Barbara Decker

Index

LAND CIRCLE

Other books by Linda Hasselstrom

Caught by One Wing
Roadkill
Windbreak
Going Over East: Reflections of a Woman Rancher

Books edited by Linda Hasselstrom

James Clyman: Journal of a Mountain Man
Next-Year Country: One Woman's View

LAND CIRCLE

Writings Collected
from the Land

Linda Hasselstrom

Fulcrum Publishing
Golden, Colorado

Library of Congress Cataloging-in-Publication Data

Hasselstrom, Linda M.
    Land circle : writings collected from the land / Linda Hasselstrom.
       p.   cm.
    ISBN 1-55591-082-3
      1. Hasselstrom, Linda M.—Homes and haunts—South Dakota.
2. Poets, American—20th century—Biography. 3. Women ranchers—
South Dakota—Biography. 4. Ranch life—South Dakota—Poetry.
5. Natural history—South Dakota. 6. Ranch life—South Dakota.
7. Nature—Poetry. I. Title.
PS3558.A7257Z467   1991
818'.5409—dc20   91-55212
                                    CIP
    ISBN 1-55591-142-0 (pbk.)

Printed in the United States of America

0 9 8 7 6 5 4 3 2 1

Fulcrum Publishing
350 Indiana Street, Suite 350
Golden, Colorado 80401-5093